THE ELECTRONIC GRAPEVINE

Rumor, Reputation, and Reporting
in the New On-Line Environment

TELECOMMUNICATIONS

A Series of Volumes Edited
by Christopher H. Sterling

THE ELECTRONIC GRAPEVINE

Rumor, Reputation, and Reporting

in the New On-Line Environment

Edited by
Diane L. Borden
George Mason University

and
Kerric Harvey
The George Washington University

LEA LAWRENCE ERLBAUM ASSOCIATES, PUBLISHERS
1998   Mahwah, New Jersey   London

The final camera copy for this work was prepared by the author, and therefore the publisher takes no responsibility for consistency or correctness of typographical style. However, this arrangement helps to make publication of this kind of scholarship possible.

Lawrence Erlbaum Associates, Inc., Publishers
10 Industrial Avenue
Mahwah, NJ 07430

Cover design by Jennifer Sterling

Library of Congress Cataloging-in-Publication Data

   The Electronic grapevine : rumor, reputation, and reporting in the new on-line environment / [edited by] Diane L. Borden, Kerric Harvey.
      p.   cm.
   Includes bibliographical references and index.
   Contents: The future of journalism in a distributed communication architecture / John E. Newhagen & Mark R. Levy — Public opinion, expert opinion, and the illusion of consensus, glenaing points of view electronically / Susanna Hornig Priest — The blurring of the line between advertising and journalism in on-line environment / Wendy S. Williams — Niche-market culture, off and on-line / Patricia Aufderheide — Cyberspace, a consensual hallucination / Jason Primuth — going on-line with the U.S. constitution, gender discussions in the cultural context of the First Amendment / Kerric Harvey — Cyberlibel, time to flame the times standard / Diane L. Borden — the campus press, a practical approach to on-line newspapers / Bruce Henderson & Jan Fernback — Technology and journalism in the electronic newsroom / L. Carol Christopher — Journalists' use of on-line technology and on-line sources / Steven S. Ross — Content analysis in an era of interactive news, assessing 21st century symbolic environments / William Evans — Making sense of the new on-line environment in the context of traditional mass communictions study / Kevin Kawamoto.
      ISBN 0-8058-2171-6 (alk. paper). — ISBN 0-8058-2172-4 (pbk.)
      1. Journalism — Data processing.   2. Internet (Computer network)
   3. Information networks.   I. Borden, Diane L.   II. Harvey, Kerric.
   III. Series : Telecommunications (Mahwah, N.J.)
   PN4784.E5E44   1997
   070.4'0285—dc21   97-27985
                                                            CIP

Books published by Lawrence Erlbaum Associates are printed on acid-free paper, and their bindings are chosen for strength and durability.

Printed in the United States of America
10   9   8   7   6   5   4   3   2   1

Contents

Preface

Journalism has always been harder than it looks. Not that it's ever really tried to pawn itself off as an easy way to make a living. Long hours, low pay, less time for lunch than most people spend doing their laundry. Granted, there's a kind of shabby glamour attached to the fourth estate, a sort of folk hero profile evoking rumpled cotton shirts, smoky bars, frayed nervous systems, and dedicated cynicism. Both on the street and in the newsroom, archetypal images of America's press corps seem to capture the quintessential traits of U.S. culture. Wise and kindly editors, brave, rumptious, and robust reporters. Rugged. Tough. Independent. Savvy. Incorruptible. Exhausted.

And everybody hates them. Since the hate is based on fear—fear of what they will say, to whom they'll say it, what they will print, what they will find out that sources didn't tell them, and what that will eventually cost—hostility is usually expressed as a twisted type of courtship. This makes conducting a private life a difficult undertaking for most journalists. The world divides into two camps. Everyone's either a source or a colleague. Either way, it also makes the touchiest part of journalism—maintaining viable working relationships with people who use them as much as they use others—a might challenging. No one likes the press, but everybody needs it. Or thinks they do.

What all those sources really need, of course, is for the press to believe them. To trust that what they say is true and to pass on to the reading (or viewing, or listening) public the "truth" as they, the source, wishes to have it told. In similar fashion, what the press needs is also a special type of faith. They need to trust in their sources, to count on the people who "tell" them the news to not stretch the facts past the point of distortion as they try to shape those facts to serve a higher agenda. Additionally, members of the press need to be able to trust themselves, to rely on their own assessments of the information they receive, retrieve, or wrestle from newsmakers and their associates. Journalists have to be able to trust their gut, to depend on the alchemy of education and experience that bubbles up from repeated exposure to news events and news-making personalities. And, because they are the unofficial scribes of humanity's public narrative, journalists must develop ways of verifying material they trust instinctively, of making a case for the obvious.

Not a small job. In the 300 or so years that we've been practicing journalism on American shores, there are still some things that we do better than others, and still some things that we don't do well at all. This doesn't necessarily reflect any sort of inadequacy or insufficiency of U.S. journalism, per se. Rather, it speaks to the almost impossible task which confronts it daily.

How gloriously audacious it is, to bring 260 million Americans "the world tonight," in half an hour or less, or to chronicle the hijinks and hilarity and sometimes outright horror that humanity visits upon itself across the globe, in whatever size newshole might be available for any given evening edition. Journalists deal in news, and news consists of the collection, verification, and recitation of facts. Where do journalists get these facts? From other information sources, if they trust them, or from actual people whom they have cultivated in a source relationship, if they believe them, or from official representatives of "newsworthy" groups, if they meet credibility standards, or sometimes from each other, if they're feeling cooperative that day and if they trust the colleague in question to have accurate information to share and no covert agenda in sharing it.

Most of these sources have some degree of physical world component to them; a trusted source is called on the phone, or met in person to talk. An expert witness is interviewed; a key document is photocopied and filed away after review. And few of these news information sources demand much information management once the initial decision—to trust or not to trust—has been made. Someone was either shot Thursday night at the 7-Eleven, or they were not. That's news. Giving some reasons why there may have been guns, people, and ill will converging in that particular store at that particular time is not, strictly speaking, news. It is news analysis, and the writing of it triggers an entirely new threshold of digging, data collection, and decision making about whose information to believe and whose to discard. Occasional demands to distinguish among facts that are true and those that are merely accurate, or among facts that are not true, but which have yielded concrete and meaningful impacts, nonetheless, clearly complicate the journalistic process. Instances where the essential character of the news story itself insists on journalists being able to make these kinds of distinctions illustrate just how tough they are to achieve. The monumental confusion surrounding the Centers for Disease Control's (CDC) first wave of press releases about how people get AIDS supplies an example of the press reporting facts that don't add up to a "truth" without a little extra help. "Facts" make no sense without the very kind of integration that is usually labeled *news analysis*.

So it's hard enough to do journalism well when you can actually get your hands on what it takes to do so, when you can read and re-read that press release, or sense that your interview subject is lying to you by the way he keeps twisting paper clips into little gnarled metal pretzels as you talk. Imagine the quantum leap in the challenges presented by traditional journalism when, suddenly, everything **physical** goes out of it. When information stops being concrete; when you're not even sure,

I

RUMOR

The Future of Journalism in a Distributed Communication Architecture

John E. Newhagen
Mark R. Levy
University of Maryland

The architectures of information technologies reflect the societal power relationships they embody. This observation is most poignant at moments of convergence, when old social systems struggle to maintain their integrity within the context of the architecture defined by the new technology (McLuhan, 1964/1994). Journalism, we contend, now finds itself at such a juncture, as it reflects on a set of mature norms and canons established during the reign of mass circulation newspapers, and as it looks ahead to computer-based information network technologies.

Newspaper and television production can be imagined as having an hour-glass shape: Large amounts of information flow in linear fashion from many sources through a narrow, journalistic "neck" and on to a mass of readers or viewers. The ability to control this linear flow rests almost exclusively with the journalist. The result is an asymmetry in *social* power, with the scales clearly tipped toward the journalist. Indeed, this inequality of power helps define the way society regards newswork and gives rise to public and professional concerns about such concepts as credibility and objectivity. Thus, for example, one key component of objectivity, *balance* in news content, becomes an issue for journalism because of the perceived *imbalance* in power between journalists and their clients.[1]

---

1. Some news sources may have the ability to co-opt this agenda-setting role by strategically positioning themselves at a point where the hourglass has already narrowed, but still prior to the journalist in the news flow. Of course, journalists are taught to be on guard against this possibility, but frequently succumb to it under daily deadline pressure. Indeed, this observation applies equally to journalists and public relations agents since both are subject to the same architectural constraints on their ability to control content.

The convergence of linear mass media technologies with nonlinear computer technologies, depicted in Figure 1.1, raises the issue of what the social relationships between information providers and receivers will look like in any hybrid technology.

At such moments of convergence, the meaning of basic concepts, which might have seemed obvious to an earlier generation, demand re-examination and explication. In that vein, we first examine the essential product of journalistic endeavor—news—and consider it as a special category of survival-enhancing information. Then, we discuss the process of news creation in the context of mass media technology. Finally, we speculate about the durability of the journalistic institution in the age of network-based information technologies.

NEWS AND SURVIVAL

Journalists report, edit, and distribute a special category of information called news. The unique character of news has to do with a basic need common to all living things for information about novel or threatening events in their surroundings (Darwin, 1872/1964). Imagine the moment when some prehistoric ancestor out on a hunt encountered a sudden rustling in the jungle bush. From an information-processing perspective, the challenge for this unkempt wag is to pay attention to the novel event, classify it based on limited information, and reach a behavioral decision about how to respond, all in real time (Newhagen & Reeves, 1992). This evolutionary imperative, manifested by the question, "Will I be its dinner, or will it be mine?" is driven by the ability to process limited information under time pressure efficiently (see Lang, 1990; Newhagen, 1994).

Across the millennia, civilization arguably has advanced, but the information-processing problem (i.e., understanding novel events in a complex environment with limited time and resources) has remained remarkably constant. If anything, the problem has become more complex because our information needs have far exceeded the range of direct sensory observation. Information-processing technologies thus perform the critical task of locating survival-specific events at distant locations, collecting raw data about them, and packaging them into messages called news. True, sometimes novel information not directly related to survival is created by those mediating technologies. A story about a frog with two heads, for example, might make the newspaper, but it is not, in the sense used here, strictly defined as news.

If news can be thought of as survival-relevant information about novel events, then the process of newswork becomes an important phenomenon for study. In what follows, we offer some preliminary thoughts about the relationship of news to the technologies that are used in its creation.

Figure 1.1
Mass media system architecture

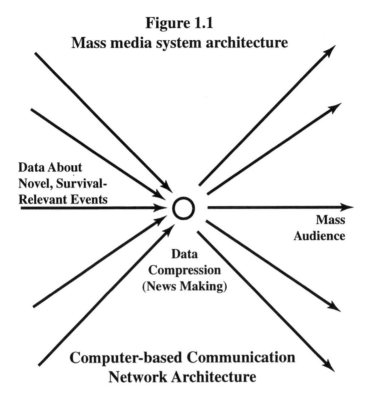

**Data About
Novel, Survival-
Relevant Events**

**Mass
Audience**

**Data
Compression
(News Making)**

**Computer-based Communication
Network Architecture**

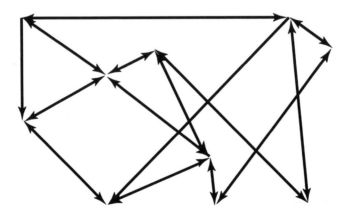

Interconnected Message Sender/Receiver Nodes

TECHNOLOGY AND THE INFORMATION SPECIALIST

Throughout history, every age has had its "defining technology" (e.g., the printing press, steam engine, or computer) and, in each of those arbitrarily delineated "ages," certain prominent characteristics or anthropomorphic qualities of the defining technologies are commonly adopted as positive societal and individual traits (Bolter, 1974). As we show later in this chapter, the social role of the information specialist is always negotiated within the context of the dominant or defining media technology at any historical moment.

In thinking about technology, we are less interested in its physical implementation and more concerned, to borrow a phrase from computer science, with its system architecture. Specifically, technology is process, with its architecture manifested in any number of physical implementations (Beniger, 1986). War drums and newspapers are both manifestations of a technology where message senders encode information into abstract symbols whose meaning is understood by their receivers. In one case, the technology, language, is implemented by using hollow logs, and in the other, printing presses are used. It is important to understand that while the messages sent by drum might contain novel survival-relevant information, they are not news unless they are the product of a mass medium system. This is illustrated by the fact that while information about novel events seems to have been valued very early in human history, the word "news" did not come into usage until the European invention of movable type by Johannes Gutenberg in the 15th century (Simpson & Weiner, 1989).[2]

While unquestionably a critical advance in human communication, neither the printing press nor movable type enabled a true mass medium. Movable type had more to do with regularity and uniformity than it did scope or, to use today's jargon, penetration (Bolter, 1991). Neither did it trigger the formation of a class of information professionals. Notice that the people who created newspapers at this point were called printers, reflecting the relationship between artisans and their press. Moreover, printers stood as peers with the merchants of the emerging middle class. As the merchants were go-betweens in the world of trade and commerce, so the printers became go-betweens in the world of information. These two observations are important, because they imply there was considerable consensus between information provider and client about content.

The word "journalism" did not come into usage until 1833 (Simpson & Weiner, 1989). That almost exactly coincides with the emergence of mass circulation newspapers in the eastern United States, a feat made possible by high-speed rotary steam presses (Emery & Emery, 1988). Those presses were able to take one message and very rapidly produce enormous numbers of exact replicas at extremely

---

2. Prior to the invention of movable type, current events were called tidings (Emery & Emery, 1988).

low cost. This added the important characteristic of scope to newspapers, and represents the implementation of a truly mass medium.

The production and consumption of a mass circulation newspaper might be likened to that of a Hershey bar. The first step in the production of the taste treat is to collect and to dry large quantities of chocolate beans at some remote tropical location, then ship them to a central production facility. At this point the raw material is concentrated into a powerful extract or essence. Only a small amount of that extract is then added to a vat of other ingredients, from which huge numbers of perfectly identical candy bars are produced and distributed.

Similarly, in news work, vast amounts of data are gathered from around the world and collected at some central production facility.[3] A similar process of concentration or compression then takes place, where data are reduced to a potent extract or essence called stories, and then combined to create huge numbers of replicas called newspapers.

Mass media architectures are sufficiently complex as to require a great deal of specialization and differentiation. The collectors have power in the process to the degree they control raw material. Their title, "reporter," suggests a relationship between them and the events they cover.[4] Editors, by contrast, derive a great deal of power based largely on their position in the message concentration process. Notice that their title (editor) refers to their relationship to the product (data/news), not to the replication technology, as was the case with the printer. Editors reside in the production facility, the newsroom, at exactly the narrowest point of information flow, where compression or condensation of the raw data takes place. Editors work at the task of data compression, creating uniform story units that enable the printing or broadcasting technology to infinitely replicate them. This power is manifested in all their editorial functions. For instance, the role of assignment editor constitutes a form of filtering Beniger (1986) calls preprocessing. This filtering role is most evident at the moment of story selection, and has been conceptualized as gate keeping and agenda setting. Copy editors, too, exert an enormous amount of power based on their ability to look within stories and actually manipulate the raw data provided by the reporters, giving it meaning by imposing structure. Given the amount of real power they have in the creation of "meaning" in the news, it's ironic that the copy editing function is depicted as one of the most tedious and least glamorous of all jobs in journalism.

Regardless of their specific function, editorial power derives from the fact that very small expenditures of energy are amplified to have enormous reach into the mass population of message receivers.

---

3. For an earlier use of this industrial metaphor to model local television news production, see Bantz, McCorkle, and Baade 1980.

4. We are fully aware of the research literature that stresses the subjective nature of news creation and the negotiated reading/viewing of news texts by audiences. However, we still contend that the power of journalists lies in their role as initial definers of newsworthiness.

In turn, the reader receives only a diluted "dose" of the concentrated data in the form of their exact replica copy. This is important to understand because, unlike earlier days of printing, in the era of mass media, message producers, particularly for the elite media, are no longer the peers of the message receivers (Harwood, 1995). Editors now control content; it is precisely their job to embed meaning or structure in the raw data (Hall, 1980). This is stated in the agenda-setting dictum that editors do not tell their readers what to think, but they do tell them what to think about (McCombs & Shaw, 1972).

The obvious need for professionalization at this stage of development can be seen by imagining what happens when a batch of chocolate extract goes bad at the candy plant. Of course, the gravity of mistake is amplified by the process of concentration, and the market ends up being flooded by bitter tasting candy bars. To avoid that catastrophe the concentration process must become highly standardized, and controlled by skilled technicians.[5]

Likewise, with the advent of mass circulation newspapers, the standard production of news became the job of a professional elite largely due to a concern with quality control. For instance, if journalists wrote each story out by hand, a factual error might be confined to just one copy of the newspaper. However, when a journalist commits an error within the architecture of a mass replication information system such as a newspaper, it will be faithfully reproduced thousands, or even millions of times. (This line of reasoning can be extended to television news production because, while it differs from newspapers in important ways, broadcast journalism is obviously still a mass media system.)

Thus, standardization and canonization help motivate the generation of "good" information. The discussion of what "good" means within the normative language of the professional revolves around the central theme of truth. The practice of techniques to insure accuracy, balance, and fairness are all deemed important to the degree they help achieve truth.[6]

While those norms lay out a practical proscriptive agenda on how to achieve the goal of objective truth, they do not really say why that goal is important. It is at this point that conceptualizing news as novel, survival-dependent information becomes valuable. Evolution has endowed the species with a set of psychological heuristics to help insure the organism attends to and processes "good" information from the proximate environment. The visual system, for instance, is designed to attend to rapid or novel motion (Gibson, 1979). Humans employ sophisticated heuristics in the assessment of risk, with outcome probabilities overweighted to emphasize

---

5. In a similar discussion of the need for standardization in complex technologies, Beniger (1986) discusses how time zones were created to help prevent railroad train wrecks.

6. These processes of standardization and canonization might also be useful in achieving the individual goals of professional autonomy and organizational goals of prestige- and revenue-enhancement (see McQuail, 1994, especially Chapter 7).

events taking place nearby (Tversky & Kahneman, 1974). Stocking and Gross (1989) discuss the role these cognitive processes play in journalistic reporting.

However, humans living in complex societies have come to need information well beyond the range of their senses. Journalistic norms, then, take on the job of psychological heuristics in generating "good" information, only at a societal level. That means an account of an event is "good" to the degree that such an account will help in making important decisions about it.[7] This can really be seen as a leap across levels of analysis, where the idea of "true" at a social level corresponds to "real" at a psychological level. Norms such as accuracy and balance are, then, heuristic responses specific to mass media. They are intended to insure that the media system generates messages that are "true" accounts about events to the degree that they can be processed by readers or viewers as if they were directly experienced and "real." The question now has to do with the kinds of psychological and social heuristics appropriate to the next generation of information systems.

THE SHAPE OF THE NET

While it is true that the Internet (or Net) may represent as significant an innovation as movable type, it is equally true that the idea of a "multimedia" personal computer system has come to be accepted with only limited understanding of its true meaning or potential long-term impact as a way to communicate. Touting "multimedia" as a feature might sell computers, but it also overlooks the important defining character of the computer as a communication technology; that is, its distributed architecture. Even within communication research, the topic has come to be identified as "computer mediated communication," which describes the technology's physical implementation, and not its architecture (e.g., see Biocca & Levy, 1995; Walther, Anderson, & Park, 1994). Similarly, if newspapers were typified by the physical implementation of the technology rather than their system architecture, we would study them as "printing press-mediated communication," rather than mass media.

However, the true power of the Internet resides in the way it is hooked up, that is, in its architecture. The flow of information through a network is nonlinear, and distributed across a vast number of sender–receiver nodes. Because message production can take place at any node in the network, information distribution is a diffuse, parallel process, unlike the compressed, serial process of mass media. This can be seen in the transmission of the simplest e-mail message. A message may

---

7. The relationship between "good" information and "making important decisions" can be considered not only in an evolutionary biological framework, but also from a liberal normative perspective. So-called "free press" theory would suggest, for example, that unbiased, objective information, delivered via unfettered journalistic mass media, is necessary to create informed citizens and to insure the most fundamental decision-making processes of a democratic polity.

become fragmented during transmission due to the packet switching protocol employed by the TCP/IP standard employed by the Internet. Thus, different parts of the message may take an entirely different route across the Net before reuniting at their destination.

McLuhan (1964/1994) argued that technologies are extensions of human processes, and the information technologies are extensions of the central nervous system. That logic makes the idea of a computer network based on a parallel distributed-processing architecture compelling. Rumelhart and McClelland (1986) describe the architecture of human learning and memory in precisely those terms. Since their watershed work first appeared, the body of evidence supporting the idea that the brain is wired into such networks has grown steadily (see Hinton & Anderson, 1989). Computer designers have even come to borrow from this psychophysiological model, constructing "neural net" programs that have the capacity to learn, and link super-computers together in "massively parallel, distributed" architectures. Neither is the similarity between the architecture of the brain and the Net lost on the so-called cyberpunk genre of science fiction, in which computer hackers "jack in" to make a seamless connection between their nervous system and the Net (see Gibson, 1984).

THE DURABILITY OF JOURNALISTIC PRACTICE ON THE INTERNET

Data concentration is unnatural in distributed network architectures that facilitate dispersed message production. Thus, the application of canons or standards produced to deal with mass media systems may be unnatural, unrealistic, and practically impossible to apply in a setting where any participant is equally likely to be a message producer as a message receiver. Members of such a system are more likely to be true peers, further eroding social codes borne out of the need to protect against the amplification of error fostered by power imbalances. First, the reportorial act of data collection is dispersed, with data collection potentially taking place at any node on the Net. Second, and most importantly, editors may lose control of the agenda.

Movable type brought standardization and regularity to symbolic human communication. The rotary steam press added scope. Networked computers now add resolution to information technology, manifested in the inherent interactivity of the architecture. Now, journalists have to examine what their role will be in the context of such parallel distributed communication architectures. An initial investigation might begin by re-examining some of the essential functions performed by journalists in a mass media system.

make democracy work more smoothly? Alas, only a "democracy" of the technologically elite is possible, and even this is a bit shaky. The great majority of U.S. citizens still don't own a computer, let alone a modem or a connection to the Internet highway, and this is greatly skewed by income. U.S. Census Bureau data for 1993 reported that 65% of households with incomes over $75,000 had computers, but only 6.5% of those with incomes under $10,000 and about one third of those in the upper middle range ($35,000–$55,000) did. Electronic discussions typically capture only a tiny minority of these; many users habitually "lurk" on the electronic sidelines, watching and listening from a virtual distance, while other, more assertive ones shape "public" discussions to their own agendas.

Of course, this is not too different from what happens in the nonvirtual world, and it needn't be all bad. If a school board calls a meeting to approve new health textbooks, chances are only a vocal minority with an ax to grind will actually speak. Self-government depends on distinguishing between the protestations of those few and the ground swell of community sentiment, but should—and does— respond to the strength of peoples' convictions as well as the numbers who hold particular opinions. It seems reasonable, many times, to argue that those who choose to attend public meetings, who have taken the time to educate themselves and form a reasoned opinion, who are members of a particular "public" with a clear stake in a particular issue, *should* have the power and the right to dictate the outcome of that issue disproportionate to their own numbers. Everyone is given the same opportunity to speak out; there's a case to be made for listening most carefully to those who have made an active effort to exercise that right.

But electronic discussions are not, and probably never will be, equally available to everyone. Public terminals in libraries will hardly solve the accessibility problem, any more than public availability of encyclopedias has solved the problem of public education. Indeed, assuring truly equal access to information—that is, ensuring the emergence of an informed citizenry—in an increasingly complex information environment is one of the unsolved problems of modern democracies. Public education itself is not a sufficient solution; witness the breakdown of the public school system in the United States and the current "science literacy" debate. Sociologists—perhaps most prominently Daniel Bell (1973)—have speculated that technologically complex societies will call into existence new forms of governance. The emergence and common use of technologically sophisticated communication systems raise parallel challenges.

THE ELECTRONIC SCIENTIST

Because the Internet was, for many years, primarily an academic researcher's system, developed with National Science Foundation and Department of Defense sponsorship for the purpose of linking researchers with similar interests who were geographically remote from one another, it's not surprising that this remains one of the system's most important, if not always most visible, uses. And as the flexibility and "reach" of Internet-based systems has expanded, science-oriented networking has exploded at least as much as any other kind. But the case of scientific consensus remains unique within the more general subject of the emergence of electronic communities. First of all, there's considerable divergence within the scientific community itself as to when, whether and to what extent scientific findings—especially preliminary scientific findings—should be publicized at all. Scientific popularizers, for a variety of reasons, risk losing status among their colleagues. According to the dominant view, popularization invites public misinterpretation and can veer dangerously close to self-promotion.

The prominent *New England Journal of Medicine* (*NEJM*) insists that research published within its pages not have been the subject of previous discussion in the popular press. This prevents the distribution of findings that have not yet been subject to the scrutiny of the peer review process, protecting the public from untested information being accepted as scientific or medical fact by the unwary press. But it also maintains *NEJM's* national prominence as a source of the latest and greatest "breakthrough" medical research. The first set of arguments may explain why research reports would not be publicized until they have been accepted for publication, but not why discussion should be suppressed until findings actually appear in print. Under circumstances like these, it is little wonder that only a minority of scientific researchers habitually appear in the media, a phenomenon Goodell (1977) tried to capture by her use of the term "visible scientist" to describe these media-friendly few.

This is a serious issue. Popularization that is too swift and insufficiently accurate has been blamed for legitimizing the preliminary claims of "cold fusion" researchers in Utah and for fueling the uninformed public fears of an earthquake predicted for New Madrid, Missouri. Popular "misrepresentations" of fact are said to be the source of public outcries over pesticide use (alar on apples), agricultural biotechnology (bovine somatotropin), and other food safety issues ("Jack In the Box" hamburgers).

Overlooked in all of these explanations, however, is the political character of science. Scientists are engaged in intense competition for recognition, prestige, and funding, and maintaining elite control over scientific accounts of truth can serve to limit public access to information needed to make good decisions about what risks people should and should not be willing to undergo. How much authoritarian control over scientific information is actually benign, whose interests are most clearly at stake here, and what role does (or *should*) popularization through the media play?

These controversies spill over into the electronic world. On the one hand, prominent scientific institutions have gleefully jumped on the electronic bandwagon, using various electronic services to make their expertise available to journalists and the general public—and to make their own in-house experts more visible in the news. On the other, dialogue among scientists, increasingly, takes place in virtual communities, a phenomenon that should be taken seriously as a factor in shaping the tone, direction, and outcome of scientific debate and the ultimate emergence of scientific consensus. But virtual communities are not exactly the same as traditional ones, and electronic debate is not exactly the same as the dialogue that has shaped science to date. The ramifications of this shift have yet to be fully explored or understood.

Scientific journals emerged from written correspondence among members of the early scientific societies. This tradition of "writing in" to these societies, to share one's own experimental results with one's colleagues, remains a vitally important part of science today. The motives behind this exchange of expert letters provided the foundation of today's scientific publishing. A key component of the scientific worldview is the belief that this kind of open sharing of information and public dialogue on its interpretation will, eventually, distill "truth." Electronic scientific dialogue seems to have emerged as a natural extension of these earlier traditions.

Less obvious is the fact that this dialogue traditionally took place within closed circles, among "gentlemen" (almost exclusively, that is among Caucasian males of upper class European origins). Opening scientific dialogue to the unqualified and uninitiated—first through popularization efforts, now through the Internet—challenges the traditional social structure of science. In the long run, this promises to extend the debate over the truth status of scientific data by opening up scientific discussions to anyone computer literate enough to figure out how to access them, scientific credentials notwithstanding. Is this a good thing, or not? Democratic access to scientific information appears to be a good thing. But this doesn't necessarily mean that anyone can "do science," nor are all scientific experts equal.

Electronic discussions have nurtured a new form of visible scientist. Even though scientists were "early adopters" of Internet as an everyday communications medium, that doesn't mean that the scientists who appear in cyberspace are a representative sample of the scientific community, any more than those who respond to an electronic poll are a representative sample of the population as a whole—or even a representative sample of the population of computer uses. And expert consensus is not the same thing as majority rule; not all opinions have equal weight, and here highly issue-specific expert qualifications must become an important criterion for the value of opinion, to be considered alongside (as for public opinion) degree of conviction and willingness to speak out on the one hand, and representativeness on the other.

Just as "public" opinion among the Internet users and lurkers ought not to be mistaken for public opinion in general, participation in Internet-based scientific discussion is highly likely to be skewed, and should not be taken as representative. Most modern-day scientists know how to send e-mail. Most of them are usually too busy to spend a lot of time chatting, however, whether electronically or in person. Therefore, it is likely that electronic discussions of scientific issues contain proportionately fewer active scientific researchers and more interested bystanders than would be found, say, at a scientific meeting. The "brakes" provided by the review and editorial procedures followed at good scientific journals, however imperfect, are almost entirely missing in the electronic world. This has a lot of potential for accelerating the rumor-formation process about scientific topics. Add to these difficulties the new opportunities for information subsidy presented to the public relations departments at universities and research centers, who are anxious to promote their experts as THE experts in a given specialty area through new on-line services, such as Profnet, that are designed to link journalists with scientific sources, and it becomes clear that the value of Internet-based scientific opinion is at least as problematic as the value of Internet-based "opinion polls."

At the same time, mainstream science has been accused, with considerable justification, of being sexist, racist, and elitist, an "old boys'" dialogue in which some explanations are arbitarily privileged over others and in which social group membership and institutional prestige may be as significant as the soundness of scientific conclusions. Scientific progress is sometimes accomplished, as Thomas Kuhn (1970) pointed out, in great leaps rather than incremental steps. But in between these great leaps (or "paradigm shifts," in Kuhn's terms), information that doesn't fit the dominant paradigm is largely ignored, especially if it is generated by people outside the traditional scientific "in-group." So to compound the problem, it is rarely safe to assume that truth is always the province of the best credentialed scientist at the most prestigious university.

Could electronic information dissemination increase the odds that a "maverick" scientific opinion that indeed turned out to explain some phenomenon better than the mainstream one would be recognized early on? Perhaps, or perhaps not. Perhaps electronic systems, which require a certain minimal level of technological expertise and equipment, might even be further biased *against* maverick views. No one, journalist or scientist, can reliably predict which scientific explanations will prove useful, and which will ultimately be dismissed as dead ends. And, of course, in the end the issue of absolute truth or falsity is, philosophically, extremely slippery in its own right.

WHAT TO DO?

Nobody—at least no thoughtful person working within the legal tradition of the United States—is about to propose "accuracy watchdogs" for the electronic information world, any more than we would tolerate the notions of governmental press councils or journalistic licensing. Basic First Amendment freedoms are at stake. But the survival of democracy itself would at least appear to depend on maintaining the integrity of our information system as the lifeblood of an informed citizenry. An educated—as well as informed—citizenry and the professionalism of individual journalists are probably still among our most important safeguards against the erosion of these traditions. Educated, informed, and professional about what? About where information comes from, whose interests it serves, what its history has been, and the extent, given all these things, to which it can be relied upon.

Just as we have striven to encourage "media literacy," we need to start recognizing that "computer literacy" (along with "science literacy") means more than the ability to punch buttons in the appropriate order. Perhaps, in all three cases, we should be talking about "information literacy" instead. **All** information comes from sources. In the case of poll data, the sources are individual respondents speaking on behalf of a larger population. Are they representative? If not, what, if anything, gives them the right to speak for those others? In case of scientific data, the sources are generally representing points of view within a broader scientific community. What gives *them* the right to speak for others? Do they represent a consensual point of view, or is the jury still out on what "consensus" on a particular issue will really turnout to be? Do the scientific opinions available represent only the mainstream view, or are there competing analyses available? All these issues are relevant to more traditional news as well, but may yet be more compelling in the world of electronic news.

Eisenstein was probably wrong in one sense. Information technologies don't need to determine our futures—unless we let them. All technologies represent opportunities and choices. The nature of these choices is sometimes apparent in retrospect; only now, for example, can we clearly see how the cult of the individual automobile in the United States might have prevented us from developing more viable means of public mass transportation. Over time, we will inevitably make choices about how our new information systems will be structured and whether they will serve the interests of democracy or oppression. Some of my more unrelentingly Marxist colleagues would be likely to respond that since it is the powerful elite who develop new technologies, it is the powerful elite whose interests will be served by them. This is an important point. But I am less fatalistic than they and would argue, in turn, that all of us will in the end contribute to these decisions in one way or another. Whether our contributions occur actively or passively is largely up to us.

REFERENCES

Bell, D. (1973). *The coming of post-industrial society: A venture in social forecasting.* New York: Basic Books.

Eisenstein, E. (1979). *The printing press as an agent of change: Communications and cultural transformations in early modern Europe.* Cambridge: Cambridge University Press.

Goodell, R. (1977). *The visible scientists.* Boston: Little, Brown.

Kuhn, T. (1970). *The structure of scientific revolutions.* Chicago: University of Chicago Press.

The Blurring of the Line Between Advertising and Journalism in the On-Line Environment

Wendy S. Williams
The American University

It is quite possible that in the next 30 years, as society moves more and more information processing onto computers, we will witness the death of objectivity as an ethical standard for the press in America. In its place, we could get a fragmented advocacy press of countless splinter publishers, representing not only political and ideological groups but also commercial interests to an unprecedented degree.

And in this new, bewildering babel of voices, it is likely that the traditional firewall between advertising and news will come tumbling down. The wall will be crushed under the weight of competition between journalists and other types of information providers; erosion will occur in the journalistic standards of objectivity, one consequence of which will be a parallel erosion in the public's ability to trust it. Opinion, marketing, advertising, information, and news will weave together so seamlessly, in the on-line environment, that the public will no longer be able to distinguish objective reporting from promotional messages.

The public already may scoff at the idea of objective journalism, as if the phrase itself were an oxymoron, but there is a special quality to news as it is produced today, at the end of the 20th century. Particularly in newspapers, but also in radio and television, the standard articulated in style guides and industry-accepted ethics codes is that news should be fair, objective, and balanced. While reporters and editors often fail to meet that standard, and public cynicism on this point runs deep, the public trust is that news should be objective.

For most journalists, one of the strongest ethical lines—critical to maintaining objective coverage—is that separating advertising from news. Reporters rarely "back off" stories that would upset advertisers, unless ordered to by their editors or publishers. Even then, some haven't, and it sometimes costs them their jobs. This is a "church/state" issue for journalists trained in the tradition of the newsroom, and yet there is no reason to assume these distinctions will exist in the online environment, particularly as objective news mixes more and more with opinion, advertising, and public relations material.

Despite this traditional separation between advertising and news, advertisers have a long and rich history of trying to influence media editorial content to their corporate advantage, both by blocking bad coverage and by trying to plant favorable stories. While most reporters and editors have strong ethical motivation to resist advertiser pressure, research has shown that publishers often allow advertisers to influence coverage in cases where there are financial pressures on the newspaper—either during times of recession, in cases of extreme competition between two newspapers, or when the newspaper is losing advertising linage because there are other outlets for advertising available within the market.

Many groups of local advertisers—real estate, auto, and entertainment advertisers in particular—have had so many problems with what they consider negative coverage of their industry in newspapers that they could very well abandon the traditional partnership between newspapers and local advertisers if there were an alternative advertising outlet that would reach as many people. Interestingly, there is no reason to assume that this partnership will automatically continue in the online environment. According to the Small Business Administration (SBA), the cost of newspaper advertising jumped 75% between 1985 and 1993, more than any other kind of advertising, including TV spots during sporting events, which rose 60%. When advertisers can find a cheaper advertising venue than newspapers, they will take it—especially if they can direct the content.

There is every reason to expect advertisers to try to continue influencing the content of news in the on-line environment. If the digital press goes through a competitive shake-out among information providers, as will probably happen at some point in its development, advertisers are likely to gain more control over content.

To understand how this will happen, however, we need to understand how both news and advertising are changing right now. The trends that have emerged in the late 20th century will shape what happens in the 21st, even if we cannot fully envision the landscape of tomorrow.

First, news itself is changing dramatically. Newspapers have suffered a serious and continuing erosion of readership in the past two decades, despite considerable efforts within the industry to be more colorful, community-sensitive and timely. Television news, as well, has undergone significant changes, brought about by commercial pressure, particularly in local coverage. While it is difficult to say what news will be like in the future, several important trends have emerged that will shape its future.

News has become extremely time-sensitive and dynamic. The night in 1991 when Bernard Shaw, reporting for CNN, stuck a microphone out the window of a hotel in Baghdad to record a U.S. bombing raid during the Persian Gulf War, changed forever the way breaking events are covered. No longer do correspondents wait to get all the facts and put them in context, neatly written. Now they phone home while they scramble and let the news anchor in Washington "interview" them about what is happening. Journalists are reporting raw and sometimes pre-verified events, which have tremendous dramatic appeal because anything can happen at any minute, as it did during the siege of the Branch Davidian complex in Waco, Texas, in 1993. Viewers now expect this type of coverage, and while they still want to read the in-depth analysis in the paper the next morning, they are less tolerant of time delays. *Time Magazine* now publishes an on-line version of itself, with daily updates of key stories, because it realizes that it will lose its readership if it cannot appear more timely. As news becomes more time-sensitive, it also becomes dynamic, changing constantly to stay current. As it becomes more dynamic, accuracy may suffer. Nonetheless, newspapers that only publish once a day may also, in this ever-changing world, ultimately prove to be too static.

News is also becoming more fragmented, partially pushed by advertisers seeking more targeted audiences, and in the on-line environment it will be able to fragment beyond imagination. Most large metropolitan dailies have gone to zoned editions so they can target news and advertising to specific neighborhoods. While large dailies are still mass media vehicles, they are under continuing pressure to fragment and localize coverage, even if that means printing several different editions of "the paper." In fact, the vast majority of daily newspapers in this country may best be regarded as specialty publications, concentrating on local coverage. A smattering of national and international news, usually provided by the wires, is thrown in as a public service. People read these papers for the coverage of local sports and local/regional political developments. At the same time, consumers continue to feed a proliferation of print and on-line special interest magazines, bulletin boards and discussion groups. More and more, "community" refers to people with shared interests. Sometimes that will be the person across the street, and sometimes that will be a stranger in a city on the other coast.

In the on-line environment, news is cheap to disseminate, while it is getting more expensive when printed on paper or produced by broadcast. In the past, the sheer cost of publishing a newspaper, or owning and operating a television station, kept news production in the hands of a relatively small number of companies and journalists. As information moves on line, many more people, groups and companies will be able to afford to tell their own version of events. Right now anyone with Internet access, a personal computer and the right software (about a $2,000 bundle) can set up his or her own "homepage" on the World Wide Web to advertise—whatever. When it is cheap enough for college students to publish their photographs and dating preferences on line, then it is time to reconsider the now truly democratic nature of the First Amendment.

In a world where everyone can be a publisher, journalists are vulnerable to losing their franchise as gatekeepers of the news. In its place, analysts predict that journalists will develop more of a guiding role, helping people find and make links among the different kinds of information they are seeking. "The more news sources you get, the more you need an editing intelligence to sort through and help you figure out what's useful," says Rob Pegoraro of *The Washington Post* (Frankel, 1995, p. 18) "Smart media will just become more valuable as the Net expands in size and diversity. If they show they know what they're doing."

Ironically, while news is fragmenting, those fragments are better able to link to other fragments in the on-line environment. In effect, news will be more connected in the future. A piece about an earthquake in Japan will link to a piece about the effects the earthquake will have on stock prices in London, as well as articles about an array of related coverage from around the world. Journalists will build the framework of links, and readers will be able to pick a trail to follow. The news is more in the layers of a story or issue, perhaps, than in breadth of coverage. Journalists may have to learn to think like librarians.

News is becoming less limited by production space. The on-line environment is endless, and television—with the advent of cable and fiber optics—virtually so. The printed newspaper's limitations of space, and linear story format, restrict the journalist's options for conveying information, compared to on-line resources.

As the technology of electronic publishing and multimedia improves, the different media used by news—text, sound, and video—are likely to become mixed together. The story of a suspected arson fire might include live video of the dramatic effort to put out the flames, a voice-text (like radio) that weaves together the basic story with live quotations, a text box analyzing the incidence of arson in the community, and a map of the damaged area and street closings. The medium used is the one best able to capture and convey the specific content. News also may become much more interactive, allowing readers a greater voice in shaping the information that is included in coverage.

The selection of news is also becoming more driven by commercial interests—not just under pressure from advertisers, but also by the companies that own the media. While not always directly influenced by specific advertising interests, competitive commercial pressures are squeezing news companies and threaten traditional journalistic standards. In his case study of several California television news stations, John H. McManus found that efforts to increase profits by expanding audience and reducing reporting costs had produced a new set of rules for the TV journalists that undermined accepted standards of newsgathering. He summed up the new rules the journalists operated under as: seeking images over ideas; seeking emotion over analysis, including avoiding complexity and dramatizing wherever possible; exaggerating to add appeal; and avoiding extensive newsgathering because it wastes time and effort, and may prove that information that makes the story attractive is actually false. Efforts to stem the hemorrhage of readership at newspapers has many papers struggling to make changes in format, presentation, content, and coverage.

As part of the trend to make news more attractive as a product, many journalists say the line between entertainment and news has become blurred—particularly in television, with the advent of tabloid news programs such as *A Current Affair*, but also in newspaper newsrooms, where reporters say they are admonished to "spin a tale"—or drop the story all together. Some analysts say that the coverage of the O.J. Simpson trial shows how journalists are even blurring the line between covering fact and fiction, giving equal weight to both tested truths and speculation. In the process, the public may forget which is which.

Simultaneously, the line between news and opinion has become blurred. Hard news itself often veers closer to opinion as reporters try to capture the essence of an event and interpret it for the public, particularly in on-the-spot television interviewing of far-flung reporters. More insidiously, journalists are now regularly interviewed as "experts" on trials and events they have covered, in effect giving them the option of writing both traditional coverage in the newspaper and then later, on television or a radio talk show, giving their opinions about the issues at hand. Not surprisingly, it is becoming more and more difficult for the public to distinguish the difference between news and opinion and the role of journalists—or to believe that journalists can operate in an objective manner.

Advertising is also changing as media technology changes. Along with news, advertising is becoming more fragmented and better targeted to individual buyers. Advertisers have progressed from targeting small communities of people with shared interests and buying habits, to collecting information about individual consumers. Some grocery chains keep a record of everything an individual buys so that they can target market particular products to them through direct mail and coupons.

Advertising is also becoming more connected with other, traditionally nonmarketing, vehicles. Movies are made not just to entertain, but are designed, from the outset, to produce a flurry of commercial tie-ins, such as children's clothing, stuffed animals and toys. There is also a trend toward more subtle advertising, with the development of identifiable product placements in television shows, such as prominent use of Adidas shoes and Speedo swimsuits in the syndicated television show *Baywatch*.

At the same time, advertising in traditional venues, such as mass market television, has become easier for consumers to avoid. "Channel surfing," which evolved out of the myriad program choices on cable, allows viewers to click past commercials. On-line viewers can click past ads just as easily, yet research shows they will browse a commercial website if there is useful information. Advertisers see that their marketing of products will do better if it is well padded with information.

Advertisers also try to mimic news, essentially giving the presentation of their product the patina of objectivity, playing off the public's belief that news is balanced and fair. Some companies make commercials designed specifically to look like news, such as using a former news anchor seated at a news desk reading a "news bulletin" about the product. If you watch closely you know it's an ad, but if you are distracted while watching (by children shrieking or by dinner boiling over on the stove) the impression left by the ad is that an independent study has found one product better than another.

And in the on-line environment, advertising, like news, can become interactive. Ads may evolve less as something readers stumble over while looking for information and more as a deliberate site consumers visit when they are shopping for a plane ticket or dinner reservations.

So why do these trends add up to a threat to the wall between advertising and news? Because the established models—the traditional ways of thinking about news, what it is and who is a journalist; the commercial models for financing news production and dissemination; and the legal framework that ensures a protected space for news—these models don't always apply in this new digital landscape.

In fact, the legal distinction between noncommercial (protected speech) and commercial speech (which the government has limited rights to regulate,) is grey at best. Commercial speech is defined as speech that proposes a commercial transaction, speech that entices you to buy a particular product. Information on the label of a can is clearly commercial because it is read as part of a purchasing decision. But what about a press release extolling the virtues of an oil company's commitment to the environment? If it is not directly linked to a purchase, it probably is not considered commercial speech and therefore the government may have little latitude to regulate it. The government can act against commercial speech that is misleading, but again, that speech is only commercial if linked to a specific purchase. Not everything that promotes a product or cause is commercial speech.

In terms of the protections of libel law, courts differentiate between news and opinion on the basis of whether statements assert facts that can be proven true or false. If they do, then it is news and can be tested for defamation. If the statements can't be proven false, then they are deemed "fair comment and criticism," and enjoy the full protection of the law. In reality, most opinion is a mix of fact and commentary. At the same time, there is no federal requirement that opinion or advertising be so labeled in newspapers. What distinctions do exist lie only in the standards held in newsrooms and in the public perception. The tiny print at the top of an "advertorial" in the newspaper that says "this is a paid advertisement" is there only because the newspaper is trying to protect the integrity of the news in the paper. The tinier the print, and the more the advertising looks like news, the better advertisers like it.

Most newspapers in America, in fact, allow their advertising departments to actually produce some sections, notoriously the real estate news section and other "special" sections designed solely to tap specific advertising dollars, such as a once-a-year section on pets and pet care. The only difference between the look of such sections and the rest of the newspaper, however, is often just a slight change in typeface. Whether the public realizes that it has moved from objective reporting to promotional material is unclear. Editors of business pages in newspapers polled by the Society of Business Editors and Writers recently said they consider special advertising sections "misleading" to the public. Particularly in the area of consumer coverage—news about new products, housing, cars, fashion, furnishings, travel, and food—objective news standards and promotional interests often mix together to produce soft news. Hard-hitting stories, such as federal reports about auto safety, are kept away from special advertising sections in the vast majority of newspapers so as not to offend advertisers. Most journalists tolerate these chinks in the firewall, but these weaknesses threaten the integrity of all news because the distinctions are easily lost on the public.

Some journalists, in fact, are beginning to notice that the meaning of the word "news" has already eroded into something different in cyberspace than in the world of print. In a recent column in *The Wall Street Journal* (Williams 1991, p. 24), Walter S. Mossberg lectured on-line readers to beware of material labeled as "news" that is instead, discussion, gossip, rumor, or uninformed speculation. "(News) isn't claims or conspiracy theories passed around by people who've done no research or reporting," says Mossberg. "And it isn't merely news releases issued by companies, government agencies, interest groups and others." Mossberg defines "real news" as the reporting of events and trends by professional journalists who aren't involved in them directly, but who have either witnessed them or interviewed the people involved.

But if a government agency, such as the U.S. Census Bureau, gives an accurate, independent account of a new population trend in a press release, why should the public eschew that in favor of the reporter's account of the Census Bureau report? In the on-line environment, readers can get both—and they may be able to get the Census Bureau report for free. The public may also trust the Census Bureau report as the more accurate account. If it is readable, why not go straight to the horse's mouth? Who is to say that won't be regarded as news by the public? And where does that leave the journalist?

Similarly, if a commercial vendor can mix subtle advertising with "information" about a problem, such as medical information about childhood ear infections, and present the package in a newslike format, at what point does the public forget that it is watching an ad? Some companies are already producing CD-ROMs that meld information with product promotion, such as Proctor & Gamble, which is experimenting with a CD-ROM ad for its Cover Girl cosmetics.

These trends already are having some remarkably stunning effects. Microsoft, the software monolith, is building a "newsroom" in Redmond, Washington, where it will produce free news that will pull in readers for the company's fee-based information services. Essentially, news becomes the bait for a company trolling for information customers. In such a climate, how will decisions be made about selection of news? Will articles be assigned—or purchased from other news vendors and repackaged—according to how well they attract specific readers? Will the company cover all the news that is fit to print? Will it cover news in only highly selected areas? Will that coverage be pro-business or independent in its coverage of commercial interests? Will it, in any traditional sense, be news? Can it advertise itself as news? And, will it link advertising or promotional material to specific "news stories?"

Pegoraro, at *The Washington Post*, says that current Net users are "culturally hostile to big business," and that Microsoft's venture will fail if it is too obviously promotional. That may be true given the population using the Internet today, but policymakers and journalists need to envision an audience far broader in background, technological capability, education and economic status than that using the Net now if they are going to develop new legal, ethical and economic models for the on-line world. The people using the Net now will be only a small portion of those using the Net in 10 years. Advertising will seek new niches in this digital world, and it will probably become even more targeted, more subtle, more informational—and more directly linked with news.

As intelligent on-line searching agents develop, most journalists fully expect readers to have a greater role in selecting the news they read in the morning. It would be simple for advertising agencies to collect that data (they would legally need the user's permission, but that is easily obtainable if linked to a cut in fees or some other commercial benefit) and then use it to target market individuals for specific types of niche advertising. If a consumer, for example, regularly selects articles about parenting, advertisers might send him a special "newsletter" about educational and health-conscious products for children, with special coupons attached to "news stories" about kids' health and development.

If a commercial vendor is producing and publishing news as a hook for consumers, it could easily link stories with ads. This is not unlike positioning an ad next to a story on a newspage, but is arguably more audacious. In an effort to provide consumers with the best possible service, should a journalist writing a story about competitive airfares provide a link through to ticket-selling websites for every airline mentioned in the story? Most journalists would howl "no," but should a reporter covering the release of a new study on alcoholism by the federal government provide a link to an on-line copy of the full report? How about links through to nonprofit groups quoted in the article? Fewer reporters would howl, but is it really any different?

If journalists become guides in this hypertext world, they can frame an issue by whom they choose to include in the road map. They will be able to link readers to other organizations, government sites, think tanks... and commercial interests. At some point the reader stops "reading the newspaper" and moves into reading promotional material, but that point may pass in a flash and barely register. The act of reading news, no longer tied to picking up the physical newspaper or turning on the television, becomes just one step in an informational quest by the consumer, one that may become harder and harder to distinguish as different from other types of information seeking.

Perhaps it is time to call in the federal government—to ask it to write a definition for the word "news," as various federal agencies have official terms for low-fat and other dietary attributes, and to ask the Federal Trade Commission to monitor misleading claims. Maybe it is time for independent news organizations to set up a national journalism licensing board. A reporter or company couldn't use the licensing trademark unless they and their journalism met certain standards. These may seem like extreme suggestions for journalists to consider, but 20 years from now they may appear more reasonable. Although, in 20 years, it may be too late.

The problem for journalists of the future may be that, in this new, fragmented world, where advertising will piggyback on news and information in ways we can only begin to imagine, there won't be a clear place for constructing a new firewall. The intriguing technological capabilities of a fully electronic world of information and news production and delivery offer us many creative possibilities. But the problem remains that the legal framework, the ethical models and the commercial methods of paying for news in the past do not translate easily into this new, multifaceted world. Advertisers have traditionally influenced news more in the deselection of things covered than in blatant unbalanced coverage in a single article or newspaper. There is every reason to be concerned about the mix and balance of coverage over time in the electronic world, particularly if news continues to fragment further to follow specific groups of readers. How overall coverage ultimately shapes public discussion, and whether that discussion will take place in a large arena with many voices or hundreds of private discussion groups of like-minded people, is at least partially the responsibility of journalists.

Overshadowing this entire discussion, as well, is the still unanswered question of who shall pay for the information highway. In a capitalist system, journalists cannot do their job if they are friendly to business anymore than if they are friendly to public officials, yet that becomes difficult when journalists are too closely linked with the commercial interests that butter their bread. The ethical standards that prop up the firewall in today's newsrooms may not get picked up by new "news providers" with other commercial agendas, and they may very well erode under competitive pressure as the world of electronic information publishing shakes out. Every act that weakens the public's belief in an objective journalistic ethic and every concession to advertisers or corporate interests poisons the ground for the new journalistic order.

SUGGESTED READINGS

Bagdikian, B. (1990). *The media monopoly.* Boston: Beacon.

Baldwin, T. F., Barrett, M., & Bates, B. (1992). Influence of cable on television news audiences, *Journalism Quarterly, 69,* pp. 651–658.

Bogart, L. (1989). *Press and public: Who reads what, when, where and why in American newspapers.* Hillsdale, NJ: Lawrence Erlbaum Associates.

Chamberlin, B. F., & Middleton, K. R. (1994). *The law of public communication* (3rd. ed.). New York: Longman.

Chao, J. (1995, June 21). Tallies of web-site browsers often deceive. *The Wall Street Journal,* p. B1.

Collins, R. K. L. (1992). *Dictating content: How advertising pressure can corrupt a free press.* Washington, DC: Center for the Study of Commercialism.

Dennis, E. E. (1986). *The media and the people.* New York: Columbia University, Gannett Center for Media Studies.

Fass, P. (1994, Fall). Media coverage blurs the line between real and unreal, fact and fiction. *Poynter Report, 5.* St. Petersburg, FL: The Poynter Institute.

Fishman, M. (1980). *Manufacturing the news.* Austin: University of Texas Press.

Frankel, M. (1995, January 22). Journalism 101. *New York Times Magazine,* p. 18.

Gans, H. J. (1979). *Deciding what's news.* New York: Pantheon.

Goldman, K. (1995, March 13). Ad industry clicks into alternative media. *The Wall Street Journal,* p. B4.

Herman, E., & Chomsky, N. (1988). *Manufacturing consent.* New York: Pantheon.

Klott, G. (1992, July). Advertisers wield clout over content, poll finds. *The Business Journalist,* Columbia, MO: Society of Business Editors and Writers.

McManus, J. H. (1994). *Market-driven journalism: Let the citizen beware?* Thousand Oaks, CA: Sage.

Meyer, P. (1987). *Ethical journalism.* New York: Longman.

Meyer, P. (1991). *The new precision journalism.* Bloomington, IN: Indiana University Press.

Mintz, M., & Taylor, S. T. (1991, October 21). A word from your friendly drug company. *The Nation,* pp. 480–484.

Shoemaker, P. J., & Reese, S. D. (1991). *Mediating the message.* New York: Longman.

Smith, A. (1980). Goodbye Gutenberg: *The newspaper revolution of the 1980's.* New York: Oxford University Press.

Warner, F. (1995, June 15). Why it's getting harder to tell the shows from the ads. *The Wall Street Journal,* p. B1.

Williams, W. S. (1991, November). Two new surveys show the industry's reach. *Washington Journalism Review*, v.13, p. 24.

Williams, W. S. (1993, August). *The effects of the 1990-1992 recession in the real estate industry on news coverage in real estate sections at five major U.S. dailies*. Paper presented at the Association for Education in Journalism & Mass Communication annual convention, Kansas City, MO.

Zachary, G. P. (1992, February 21). Many journalists see a growing reluctance to criticize advertisers: They say some newspapers, suffering tough times, are softening coverage. *The Wall Street Journal*, p. A1.

Niche-Market Culture, Off and On Line

Patricia Aufderheide
The American University

How will on-line journalism affect public life in a pluralistic, democratic nation? Will we see enterprising citizens exploiting their newly found "infowealth" to make decisions, and searching out electronic allies to form grassroots coalitions that open up the corridors of power to the people's voice? Or does the do-it-yourself model of "infoharvesting" foreshadow a world in which consumers are pegged into ever tighter market niches, one in which they seek out only other like-minded souls and fall victim to con games that take advantage of their accessibility on the Net?

The very way the question is posed—and it's posed a lot these days, as our scenarios reconfigure every 6 months—bespeaks our national love affair with technological determinism (Iacono & Kling, 1995). We love to believe that new technologies will magically transform our societies, whether for better or for worse (Huber, 1994; Smith & Marx, 1994). In 1944, *Scientific American* foresaw that television "offers the soundest basis for world peace that has yet been presented. Peace must be created on the bulwark of understanding. International television will knit together the peoples of the world in bonds of mutual respect; its possibilities are vast indeed" ("Fifty years ago today," 1994).

But as communications scholar Jennifer Daryl Slack reminds us, possibility and probability are not always closely related. Technological practices "are as much political, economic, cultural, environmental, and ethical practices as they are strictly speaking technical" (Slack, 1994). So if we want to speculate about the relationship of niche marketing of information to public life, we do not need to wait until people can call up their own, do-it-yourself newspapers on magic screens (Fidler, 1991), or until they entrust their personal intelligent agents to stock their own customized Net niches. We can look at the world we live in today, and ask ourselves what the quality of our public life and our democratic practices are now. If we like them now, we'll love them on line.

Yes, on-line journalism inevitably restructures relationships, redefining who is a client and what is a product. And yes, there are undoubtedly enormous changes ahead. But they will result not from the power of new technologies in themselves but from their deployment within well-entrenched economic and cultural patterns. Therefore, the problems and opportunities of tomorrow will bear at least a family relationship to the problems and opportunities of today.

The challenge for journalists who confront electronically restructured markets and social relationships is one that ought to sound familiar. It is the challenge of participating in and fomenting democracy in an increasingly globalized and image-oriented commercial culture. It is also, in service of that goal, the sculpting of professional standards that insist on the journalist as a creator and defender of democratic public space.

OPTIMISM, PARANOIA, AND ANXIETY

Journalists already feel acutely the social weight of rapid changes in technology. Scenarios range from the utopian to the paranoid. The 1994 Nieman conference, "Can Journalists Shape the New Technologies?" (Kovach, 1994), deftly and authoritatively showcased the turmoil in the field.

News professionals have, of course, immediate and parochial anxieties. Journalists—who have barely won their professional status—are now threatened by the idea that their gatekeeping function may be pre-empted by ordinary readers and viewers. Ed Fouhy, director of the Pew Center for Civic Journalism and the organizer of 1992 presidential debates in which candidates spoke directly with audiences as if they were talk show hosts, notes, "The journalistic model is top down, and the whole lesson of what we're seeing now, made possible by technology, is that the 'gatekeeper' function is being tremendously weakened" (personal communication, July 22, 1994).

But there are larger and more serious issues involved than turf-guarding, most centrally the future of journalism's historical mandate. Journalists and scholars of media see an ongoing and accelerating shift in the social role of journalism, which has been to provide communities with a public space—a virtual meeting place, with a common body of daily knowledge for the citizenry (Rosen, 1992). Giving ground on this claim means, for journalists, acceding to the role of publisher's pawn or adman's tout. It certainly furthers the transformation of the citizen into an infoconsumer. That infoconsumer is, according to some analysts, not the choice-empowered individual of advertising fame, but an artifact of the desires of large corporate and bureaucratic systems, which pursue the unwary credit card holder for their own ends (Gandy, 1993; Schiller, 1989).

In one sense, the info-future of finely sliced and diced demographics is now. For the elite who now have access to the World Wide Web, electronic services that shop for interests and audiences have debuted and get more elaborate daily with the introduction of programming language "Java," which allows users to download not just data but whole mini-applications. Less fancy marketing and promotional services are multiplying throughout the Net. But that should not be surprising. In other senses as well, the future is now.

INFORMATION AGE, MARKETING AGE

The problem of creating and maintaining public space in a commercially oriented democracy is not new or emerging, any more than is the division between information haves and have-nots. The balkanization of the American public into demographic nuggets organized by zip code or car preference, the erosion of media space for noncommercial and public uses, and the segregation of the infohaves and have-nots on the basis of creditworthiness are well-entrenched processes and have reshaped the texture of American culture.

An *Advertising Age* ad demonstrates succinctly the social deployment of emerging technology. Accompanying a picture of a smiley face on a computer screen, the copy reads:

> You're not falling for this, are you? The cute little screen icons. The flashy graphics of online services. Magazines and newspapers getting on the Net to snag new readers. Those sleek, ultra-designed laptops. The 'Information Age,' is it? That's funny. It looks like...sounds like...tastes like...marketing. And you should know. That's what you do. In fact, you've had this sussed out for a while now, haven't you? It's like TV in the '40s, like radio in the '20s; another really good way to sell stuff....
>
> After all, the Information Age is really the Marketing Age. Which is why you should be in Ad Age.

Of course, it's just not that simple. Every medium presents new marketing challenges, as a public relations executive recently noted (Ellis, 1995). Nonetheless, the new possibilities this public relations pro finds in the Net sound distinctly like refinements on existing techniques: cultivate people with an interest in common, and attract users to your location with something compelling.

If *Advertising Age*'s ad is right, then we have been living in the Information Age for some time now. The most powerful revenue-generating mechanism in U.S. mass media is advertising, which shapes entire industries (Bogart, 1991; Gomery, 1993) and the nature of their editorial products. Of course, advertisers sometimes directly jump the imaginary firewall between the editorial and business sides of mass media; this is widely regarded as bad behavior. But far more profoundly, they shape the very structure of the business, whether in electronic media or in magazines (Bagdikian, 1987; Collins, 1992). They reward the efficient harvesting of particular demographic segments.

And so two media trends have resulted. Editorial content, inevitably and increasingly, seduces rather than informs, encouraging the restless fun-seeking that is also its major challenge. "Clutter," "info-overload," "zapping"—they are all buzzwords that point to the uncommitted, dislocated consumer. Second, media products and services harvest ever-more precisely sculpted audiences, selecting exactly those whom advertisers want to reach with minimal waste—"waste" being nonconsumers.

Neither trend has been hospitable to a journalistic mandate. What *Rolling Stone* dubs the "New News"—the putatively more egalitarian combination of talk shows, tabloid media, calls-ins, and ads that have changed the face of electoral reporting (Taylor, 1992)—devolves from fundamental changes in the way big media do business. Forums in most mass media that might offer a range of opinions about public affairs and an opportunity for public discussion are few and diminishing. In magazines, the "dinosaur" general weeklies are on the ropes; *Harper's* has to be bolstered by foundation funding; public television is perpetually embattled and grows ever more commercial; all politically partisan publications, from left to right, are subsidized. The magazines that do well, like *Women and Guns* (a runaway success story among niche-market magazines), discover new "taste publics" rather than provide new public platforms.

Broadcast television's small space for news and public affairs steers away from investigative and beat reporting, and toward entertainment and titillation. The local evening news has long since opted above all for consumer-friendliness, captured in the term that TV anchor Fred Graham gave an acid edge, "happy talk" (Graham, 1990). Daytime is wall-to-walled with tabloid TV (Ricki, Leeza, Gordon, and others), where the otherwise invisible lower class becomes a freak show scantily disguised as a discussion of important social issues (transvestism, anorexia, parental neglect). The night schedule is larded with "reality" programming such as *Cops, Hard Copy,* and their descendents, which make class contempt a kind of blood sport and further erode a sense of virtual place by destabilizing the very meaning of appearances (Fiske & Glynn, 1995; Nichols, 1994). The subjects and the packaging of scandal, notoriously controversial as boundary-crossing phenomena in popular culture, may shift and change with pressure; but the sale of packaged emotion is a constant (Spain, 1996).

Commercial radio's talk shows mostly masquerade as opinion and debate, instead selling the smugness of conviction and the self-righteousness of self-assertion. Rush Limbaugh only accepts calls from abject fans, or "dittoheads." But there are a lot of places to stop on the dial. Once there, listeners are likely to hear what they already agree with. It is interesting that bad-attitude male personalities like Don Imus (who describes himself as a "person who is basically pretty immature") and that master of profitably bad taste, Howard Stern, draw huge audiences. They sell, with their behavior, the thrill of being antisocial with impunity. They thumb their noses at the very concept of playing a public role.

The niche-marketing of America is already a fact, in media as elsewhere. It also has produced a host of infoproducts and services—catalogues, for instance, and print, audio, and video newsletters—tailored to the instrumental needs of a particular virtual community, whether of paleobotanists or trolley car enthusiasts. These services produce entire new mini-markets, a fact not lost on new-era marketers. Nor are new communications technologies likely to shift the trend. The Net is now being envisioned as a place where content of all kinds becomes the bait to establish ongoing relationships with consumers. "The challenge for advertisers is to make sure that their advertising messages are inextricable from the content that surrounds them," writes computer guru Esther Dyson (1995, p.182).

Daily life is already mass-mediated to an unprecedented degree. Commercial mass media appear to roam ceaselessly for new frontiers; CNN's Travel Channel in airports and the proposal for television screens (to allow personalized instant replays) on each seat of Washington, DC's, new sports stadium are two examples. While "choice" is endlessly celebrated, it is consumer choice—its parameters preselected by marketers—that is delivered. The niche-marketing of news, for example, is exemplified by MTV news' focus on bulletins about the touring schedules of the hottest musical groups.

The proliferation of "infotainment" and of highly targeted information is not merely an economic fact, but a cultural reality. There is, manifestly, a broad appetite for information that caters to lifestyle, hobby, and entertainment concerns. This is one indication of the way in which daily life has become a never-ending process of making consumer choices about relationships and activities that define social location (Aufderheide, 1986; Giddens, 1991; Phelan, 1995). Concepts of community and social identity appear to be shifting, in a commercially oriented, increasingly globalized mass culture (Harvey, 1990, pp.284–323; Lash & Urry, 1994, pp.31–59). Participation in civic and voluntary associations continues to decline, as it has over the last two decades. Cynicism about traditional, institutional politics is at an all-time high. So is fascination with issues of social identity—Afrocentrism, political correctness controversies, multiculturalism debates, controversy over lurid and violent content in media, and emerging rights movements such as disability rights and proliferating gender rights all being examples. These shifts inevitably affect what is meant by politics and by public life.

JOURNALISM AND PUBLIC LIFE IN THE MARKETING AGE

Media now and increasingly cater to demographic clusters that are appetizing to advertisers or to highly defined interest groups, not to physical communities or to publics in the sense that Habermas (1989) or John Dewey (1983) might have meant. While espousing very different arguments about the revitalization of civic life, both philosophers argue that publics are self-constituted, autonomous social groups, which can independently address the implications and consequences of other networks of power on the quality of shared daily life. In order for publics to act as publics, people have to see themselves as part of that larger whole, that autonomous social creation. By contrast, people organized into demographic clusters do not necessarily see themselves as part of the same universe as someone from a different cluster, even though both may face the same challenges from, say, a proposed development down the road. Indeed, individuals may themselves perform as different selves in the many different contexts through which they navigate and within which they unceasingly position themselves.

Daily newspapers, which typically live on very local advertising and sales, have historically had a structure distinct from other mass media. They have offered a product pitched at an entire physical community, providing not just information about, but a forum within communities that are larger, denser and more complex than any individual's ability to experience directly. But newspaper publishers have noted with alarm a steady shrinkage in advertising parallel with a decline in readership, focused particularly on younger people, who don't seem to be returning as they age (Denton & Kurtz, 1993; Ungaro, 1991). Perceptive newspaper editors have noted that this decline appears related to changing conceptions among readers of what they need to know to make it through their days, what they think are significant relationships in their lives and what kind of role they see themselves playing in their world. Many of the communities that are intensely real to people now are virtual, and many of those communities are formed around consumer decision making.

One contingent of newspaper executives has struggled to redefine the newspaper's mix of information to be more entertaining and demographically targeted. Another and much smaller contingent has grappled with the challenge of using the newspaper's resources to create public spaces in community life, by sponsoring events, involving readers in political process, and even organizing long-term urban planning projects (Austin, 1994; Rosen, 1992).

These concerns among newspaper publishers and editors—for their survival, as communications services dependent on a multilayered concept of community—are not driven primarily by technology. They are affected by cultural movements that were already vividly apparent decades before, shaping a commercial culture of consumption (Lears, 1994; Sussman, 1984, e.g., pp.211–229). Their implications for democratic life have been discussed widely (e.g., Abramson, Arterton, & Orren, 1988; Bagdikian, 1987; Bennett, 1992; Dionne, 1991; Entman, 1989; Harvey, 1990; Jacobson & Mazur, 1995; Postman, 1985), from voting to community organization to attitudes toward politicians and public life.

Newspaper editors are thus on the front lines of journalists' struggle with changing definitions of what is public, and what is community.

MEGAMEDIA AND THE JOURNALISTIC MANDATE

Content creation, manipulation, and tailoring to mini-demographics is critical to the development of widely popular new electronic services. The much-touted consumer choice idea depends heavily on having something from which to choose. In response to the challenges of a rapidly changing media marketplace, the largest communications firms have chosen the strategy of centralization and vertical integration. Starting in the late 1970s, deregulatory fervor, combined with technological innovation, dismantled longstanding industry arrangements (Horwitz, 1989). The phone company became a set of businesses hungry to enter the television programming and data delivery businesses; cable TV companies became rivals for basic phone service; several new television networks emerged (most importantly Fox); broadcasters began to lust after pager service business; Hollywood studios became part of international conglomerate deal-making. Vertical integration and cross-ownership became the rule among the largest industry players (Bagdikian, 1987; Miller, 1996, pp.144–151). Newspaper chain owners such as Gannett, Cox, Tribune, and Hearst also racked up purchases of broadcast stations and magazines, and put in motion plans to cross-market their digitized product across media. Computer companies have become partners with phone companies and movie studios to produce both distribution systems and content. Phone companies in particular have become the newest, biggest, and most ambitious aspirants to total communications provision and service (Miller, 1996, p.147), to the title of megamedia (Maney, 1995).

If content and conduit are now inextricable, the importance of content remains central. The linking of content and distribution has driven both mergers and alliances recently. Several large multinational corporations—News Corp., Time Warner, Sony, Viacom, Disney/ABC-Cap Cities—have pioneered centralized culture brokering. Their interests in civic and public life have been as minimal as the profits in those areas. Sites of public-space journalism such as broadcast news and upscale book publishing have been made into profit centers, drastically reshifting priorities and shrinking their public mandate (Auletta, 1991).

The lords of these domains have not been shy to exert their influence throughout their organizations. For instance, the trade paper of the movie business, *Variety*, discovered a trend in recent scandals involving the industry. Authors who wrote about entertainment-industry subjects or people involved with their publishers' corporation have repeatedly found their book deals dropped ("Conglom fever," 1993). *TV Guide* is notorious for showcasing Fox productions (which its publisher, Rupert Murdoch, also owns). The flamboyant Ted Turner has treated his news service, CNN, as a platform for his personal opinions (Goldberg & Goldberg, 1995, p.288f). NBC, owned by General Electric, was caught slanting news about GE's shabby defense contracting practices (Collins, 1992, p.28). When Disney bought ABC-Cap Cities, it promptly dumped the editor of ABC-owned *Los Angeles Magazine*—Robert Sam Anson, who was writing a Hollywood book on Disney head Michael Eisner (Lieberman, 1996). As vertical integration has proceeded, each of the networks has made it common practice to feature stories about its own film and television productions on its own news programs.

It is now possible for a national politician to normalize the all-too-prevalent practice of shaping editorial product around corporate agendas publicly, without reproof. As speaker of the House Newt Gingrich, speaking to a reporter for the trade magazine *Broadcasting & Cable*, said with apparent indignation in 1995:

> The business side of the broadcast industry ought to educate the editorial writing side of the broadcast industry. I mean, I went into a major cable company that owns a daily newspaper and the newspaper's editorial page is attacking the very position of the cable company. I think the managers ought to sit down in a room with their writers and talk through market economics. (Soundbite, 1995)

Just as important as—perhaps even more important than—the eroding line between business and journalistic priorities is the nature of the business. Vertically integrated info-companies—companies with theme parks to fill and TV series to develop and lunch-box licensing to reap—benefit from cross-marketing strategies that begin with conceptualizing the product. Each item gets weighed according to its marketing potential. When *USA Today* reporter David Lieberman interviewed Michael Eisner on the occasion of the Disney/ABC-Cap Cities merger, Eisner enthusiastically celebrated the value of ABC news shows—as "healthy, adolescent brands" (Lieberman, 1996).

The megamonoliths that now dominate the media landscape have the same objectives for on-line services as they do for the rest of their operations: maximum profits. For instance, Murdoch aspires to on-line gaming, based on newspaper data, such as stock market information (Rohm, 1995, p.39). Home shopping and video on demand are other hot prospects within megamedia. The information generated by journalists is part of a package of resources and assets.

Many see the blossoming of the decentralized Net, and the emerging software that allows for creation of virtual publishing, as a way around corporate control of culture. Leaving aside fond fantasies of future technologies, this hope needs to be tempered by experience. Established Net information services so far suggest the powerful role of brand named information and of corporate alliance, and a reinforcement of the have/have-not gap. For example, as Net scholar Rob Kling points out (1995), customers using high-priced data retrieval services on Dialog or low-priced ones on Compuserve or America Online (AOL) are not going to discover any of the small opinion magazines that now occupy boutique positions in publishing. Rather, they encounter the major, mainstream news sources—wire services, elite newspapers, mass distribution magazines. He also cites hand-in-glove arrangements, such as the magazine *Wired*'s running of a highly favorable article on AOL, which carries the magazine on line.

Vertical integration in the area of corporate information production has the power to shape fundamentally the uses of new communications technologies. It permits what professional journalists see as corruption and abuse to be built into the very creation of projects. It erodes the ideological space for journalism's public mandate, judging all information and communication by its ability to contribute to corporate synergy. The process will accelerate further. The Telecommunications Act of 1996 endorsed vertical integration in media industries on a record scale—a phenomenon that can only partly be explained by millions of dollars in campaign and political contributions by the megamedia (Auletta, 1995). At the same time, legislators (backed by powerful industry lobbyists) decisively rejected provisions for nonprofit access, as well as funds for educational and other nonprofit experimentation with electronic infonetworking—ways to reduce the gap between "infohaves and have-nots" and help make communications networks accessible to the entire society (Miller, 1996, pp.129–131 and passim).

Perhaps the most significant aspect of the creation of these megamedia in the late 1980s and 1990s has been the muffled quality of public debate about their consequences. Structural critique—analysis of ownership and influence—has largely issued from the powerless left press. Mainstream debate, including in Congress, has focused on the familiar terrain of sex and violence in particular products, a discussion focusing on the values issues that are one hallmark of lifestyle politics.

The market convergence going on in media industries today, boldly manifested by the aggressive vertical integration of the largest companies, accompanies technological convergence. But the relationship between market and technical convergence is a complicated one, driven by the passions of the powerful to get and maintain position. Therefore, the social effects of on-line journalism will also reflect that intertwining of socioeconomic and technological realities.

TECHNOLOGIES OF ABUNDANCE

Experience with earlier technologies also can be some guide here. The most recent claim of consumer sovereignty and social revolution on a technological basis was cable, the "technology of abundance" (Shenk, 1995; Streeter, 1987). Pundits waxed eloquent about its possibilities: "The educational and social impact of cable technology is likely to be greater than that of any other forseeable advance in telecommunications technology" (Gillespie, 1975, p.1). With so many channels, cable companies could easily relinquish some to the public, where access services might "revolutionize the communication patterns of service organizations, consumer groups, and political parties, and could provide an entirely new forum for neighborhood dialogue and artistic expression" (p.3).

But cable companies were permitted to build their business on a broadcaster/ editor model, not on a common carrier model. They constructed networks that centralized control. At the moment, consumers of cable have an abundance of choices, but among remarkably similar kinds of products, many of them strikingly, ambitiously vulgar. In perhaps a quarter of the cable systems in the nation, viewers have the opportunity to be producers at public access studios—most of them woefully underfunded. But many citizens do not avail themselves of that opportunity, except when access center directors act as community organizers. In that case, it is not the technological innovation in itself that makes the difference, but the fact that it is deployed as part of an agenda to shift power relationships (Aufderheide, 1992).

Tomorrow's communications networks could and still might look something like today's phone systems—that is, equally accessible at any point in the system, whether a teenager's bedroom or the phone company's office. If, however, tomorrow's news shoppers use a service that is organized more like today's cable—powerful output from a centralized source, limited feedback from customers—they are nothing more than audiences, albeit ever-more-finely sliced ones.

We already know which model megamedia prefer, as they explore the uncharted new world of competitive telecommunications: cable TV (Miller, 1996, p.148f). Its abundance of choice lies principally with the owners, not the consumers, of media and communications services. Megamedia owners have perked up their ears at a new media category created by the Telecommunications Act: an Open Video System. A cable TV operation, whether run by a newspaper, a cable company, or a phone company or some hybrid, would be free of cable regulation if two thirds of its service were open to any comers. This system design offers owners a hefty chunk of customers' choices, without requiring that viable competition for that "open" part of the system even has to exist.

As well, we know the general preferences of industry leaders in terms of open technology, which would allow linking of networks and thus widespread citizen access to information. The lust of major corporations such as TCI and Microsoft for proprietary technology—which would limit openness—is an indication of their ceaseless search for control of the market. So is the recent case of Austin, Texas, where the city government has proposed building an open, broadband communications network. Southwestern Bell and infoconglom Time Warner's cable subsidiary there have both vigorously opposed the service, on which they could rent space. They prefer to own systems they control, systems that almost assuredly will not be fully interactive (Chapman, 1995). Phone companies have been disappointed with their demonstration projects in interactivity, which has led to exploring more centralized, less open but cheaper systems.

On-line journalistic services could enhance the efficiency, elegance and utility of all kinds of information for an already niche-marketed society. But they will not necessarily fuel the radical vision of technological utopians such as the Tofflers and George Gilder, who imagine electronic networking as the "knitting together" of the "diverse communities of tomorrow, facilitating the creation of 'electronic neighborhoods' bound together not by geography but by shared interests" ("Cyberspace and the American dream," 1994, p.6).

Even before on-line virtuality, marketers had pioneered the cultivation of diverse virtual communities of interest, using technologies as old-fashioned as the postal service (for direct mail and newsletters) and the telephone. And they have aggressively shaped the culture of American virtual communities, so that shared interests tend to run along the lines that marketers encourage. It is easy to imagine the role of on-line journalism in fomenting this existing trend. It is much harder to imagine how journalism in an electronic age can cultivate a sense of community, of shared problems, of the need for knowledge, of respect for and curiosity about difference.

JOURNALISM AND THE ELECTRONIC PUBLIC SPHERE

Journalists who prize the social role of journalism in a democracy and within a pluralistic culture have plenty of work to do, as we hurtle into the 21st century. But the challenge of new technologies, and in particular networked communication, is only one piece of a much larger puzzle. The larger question is the fostering and reproduction of democratic culture in the Age of Marketing. The awareness of some newspaper editors and publishers that the very notion of public life is at stake is a healthy beginning, one of many possible ones.

The rapid changes in journalism do put into focus some basic questions of professional behavior, as Jay Rosen has argued so eloquently (Rosen, 1993, p.26f). Journalists need to grasp the significance of their own profession, a significance that only becomes more important as information becomes a malleable piece of recombinant commercial culture. They need to put onto their own news agendas— their own sets of curiosities—the very structure of the information and communication industries. Then they can fight to create places and ways to incite public discussion of public television, or set-asides for nonprofit use of bandwidth, or cable access, or access to communications for schools and hospitals and libraries for what they are—struggles over the raw materials of our communications systems and the vehicles for an electronic-era culture.

Journalists also need to act as empathetic but not sycophantic ethnographers of cultural pluralism and daily democracy. The values and lifestyle controversies of today are not frivolous. They are genuine manifestations of majestic pressures on individuals in highly fluid cultures. Feature writing and lifestyle issues have become front page material for legitimate reasons as well as because of consumer-centered marketing strategies.

And finally, journalists need to see themselves as the facilitators of responsible public discussion, not the guardians of public knowledge. They need to be the people who help us to make the connections between pieces of information that we are too busy or harried or ignorant to make for ourselves. Whether they do that by hyperlink or snail mail doesn't change the basic task, which does not get any easier with new technologies but just might be done creatively and well with them.

REFERENCES

Abramson, J. B., Arterton, F. C., & Orren, G. R. (1988). *The electronic commonwealth: The impact of new media technologies on democratic politics.* New York: Basic Books.

Aufderheide, P. (1986). The look of the sound: MTV. In T. Gitlin (Ed.), *Watching television.* (pp.111–135). New York: Pantheon.

Aufderheide, P. (1992). Cable television and the public interest. *Journal of Communication, 42*(1), 52–65.

Auletta, K. (1991). *Three blind mice: How the TV networks lost their way.* New York: Random House.

Auletta, K. (1995, June 5). Pay per views. *New Yorker,* 52–56.

Austin, L. (1994). *Public life and the press: A progress report.* New York: Project on public life and the press, Department of Journalism, New York University.

Bagdikian, B. (1987). *The media monopoly* (2nd ed.) Boston: Beacon Press.

Bennett, W. L. (1992). *The governing crisis.* New York: St. Martin's.

Bogart, L. (1991). The American media system and its commercial culture. *Media Studies Journal, 5,* 13–34.

Chapman, G. (1995, November 30). Op-ed, Los Angeles Times, accessed via chapman@mail.utexas.edu.

Collins, R.K.L. (1992). *Dictating content? How advertising pressure can corrupt a free press.* Washington, DC: Center for the Study of Commercialism.

Conglom fever plagues book biz. (1993, October 11). *Variety*, p. 10.

Cyberspace and the American dream: A Magna Carta for the Knowledge Age (1994, August). Washington, DC: The Progress and Freedom Foundation.

Denton, F., & Kurtz, H. (1993). *Reinventing the newspaper.* New York: Twentieth Century Fund.

Dewey, J. (1983). *The public and its problems.* Athens, Ohio: Swallow Press. (Original work published 1927).

Dionne, E. J. (1991). *Why Americans hate politics.* New York: Simon & Schuster.

Dyson, E. (1995, July). Intellectual value. *Wired,* 3.07, 135f.

Ellis, L. (1995). *Communicating in chaos: Corporate presence in the online world.* New York: Fleishman-Hillard.

Entman, R. (1989). *Democracy without citizens: Media and the decay of American politics.* New York: Oxford University Press.

Fidler, R. (1991). Newspapers II: Mediamorphosis, or the transformation of newspapers into a new medium. *Media Studies Journal,* 115-126.

Fifty years ago today. (1994, June). *Scientific American.*

Fiske, J. & Glynn, K. (October, 1995). Trials of the post-modern. *Cultural studies,* 9(3), 505-521.

Gandy, O. (1993). *The panoptic sort: A political economy of personal information.* Boulder: Westview Press, 1993.

Giddens, A. (1991). *Modernity and self-identity: Self and society in the late modern age.* Stanford: Stanford University Press.

Gillespie, G. (1975). *Public access cable television in the United States and Canada.* New York: Praeger.

Goldberg, R., & Goldberg, G. J. (1995). *Citizen Turner: The wild rise of an American tycoon.* New York: Harcourt Brace.

Gomery, D. (1993). Who owns the media? In A. Alexander, J. Owers, & R. Carveth, Eds., *Media economics: Theory and practice.* (pp.47–70). Hillsdale, NJ: Lawrence Erlbaum Associates.

Graham, F. (1990). *Happy talk: Confessions of a TV newsman.* New York: Norton.

Habermas, J. (1989). *The structural transformation of the public sphere: An inquiry into a category of bourgeois society.* (Thomas Burger and Frederick Lawrence, Trans.) Cambridge: MIT Press. (Original work published 1962.)

Harvey, D. (1990). *The condition of postmodernity.* Cambridge, MA: Blackwell.

Horwitz, R. B. (1989). The irony of regulatory reform: The deregulation of American telecommunications. New York: Oxford.

Huber, P. (1994). *Orwell's revenge: The* 1984 *palimpsest.* New York: Free Press.

Iacono, S., & Kling, R. (1995). Computerization movements and tales of technological utopianism, In R. Kling (Ed.), *Computerization and controversy: Value conflicts and social choices, 2nd ed.* New York: Academic Press.

Jacobson, M. F., & Mazur, L. A. (1995). *Marketing madness: A survival guide for a consumer society.* Boulder: Westview Press.

Kling, R. (1995). Boutique and mass media markets, Intermediation, and the costs of on-line services. *The Communication Review,* forthcoming. (Accessed on http://www.ics.uci.edu/~kling.)

Kovach, B. (Ed.). (1994, Summer). "Can journalists shape the new technologies? Toward a new journalists' agenda: Responding to emerging technological and economic realities—a Neiman Conference," *Neiman Reports, 48*(2), 3–73.

Lash, S. & Urry, J. (1994). *Economies of signs & space.* Thousand Oaks, CA: Sage.

Lears, J. (1994). *Fables of abundance: A cultural history of advertising in America.* New York: Basic.

Lieberman, D. (1996, February). Carlos McClatchy lecture on media and journalism, Stanford University, Stanford, CA.

Maney, K. (1995). *Megamedia shakeout: The inside story of the leaders and the losers in the exploding communications industry.* New York: Wiley.

Miller, S. (1996). *Civilizing cyberspace: Policy, power, and the information superhighway.* Reading, MA: Addison-Wesley.

Nichols, B. (1994). At the limits of reality (TV). In B. Nichols (Ed.), *Blurred boundaries: Questions of meaning in contemporary culture.* (pp. 43–62) Bloomington: Indiana University Press.

Phelan, J. (1995). *People like you: Casting for the multicultural market.* Donald McGannon Communication Research Center, Critical Studies Paper number 1. New York: Fordham University.

Postman, N. (1985). *Amusing ourselves to death: Public discourse in the age of show business.* New York: Viking.

Rohm, W. G. (1995). Rupert Murdoch: Global media mogul. *Upside,* p. 39f.

Rosen, J. (1992, Winter). Forming and informing the public. *Kettering Review,* 60-70.

Schiller, H. I. (1989). *Culture, Inc: The corporate takeover of public expression.* New York: Oxford University Press.

Shenk, J. (1995, June). The robber barons of the information highway. *Washington Monthly 27*(6), 17–22.

Slack, J. D. (1994, Nov.–Dec.). The university's technology policy. *Academe,* 37–41.

Smith, M. R., & Marx, L. (Eds.). (1994). *Does technology drive history? The dilemma of technological determinism.* Cambridge: MIT Press.

Soundbite. (1995, May/June). *Columbia Journalism Review,* p. 22.

Spain, W. (1996, January 15). Talk shows heed loud dissension from new voices. *Advertising Age,* p. 28.

Streeter, T. (1987). The cable fable revisited: Discourse, policy, and the making of cable television. *Critical Studies in Mass Communication, 4,* pp.174–200.

Sussman, W. (1984). *Culture as history: The transformation of American society in the 20th century.* New York: Pantheon.

Taylor, P. (1992). Political coverage in the 1990s: Teaching the old news new tricks. In J. Rosen & P. Taylor (Eds.), *The new news v. the old news: The press and politics in the 1990s* (pp. 37–69). New York: Twentieth Century Fund.

Ungaro, J. (1991). Newspapers I: First the bad news. *Media Studies Journal, 5,* 101–114.

II

REPUTATION

5

Cyberspace:
A Consensual Hallucination

Jason Primuth
Channel 4000, Minneapolis

> *Cyberspace. A consensual hallucination experienced daily by billions of legitimate operators in every nation.* (Gibson, 1984, p. 51)

Virtual reality, virtual banking, virtual parenting are phrases increasingly thrown around. Obviously, they employ new technology. But what else? Is their impact merely technological? More specifically, what does the increasing trend toward virtual experiences do to our notion of community?

The number of virtual communities seems to be exploding. Perhaps more significant, these communities are no longer limited to the computer fanatics who established them. Instead, average, nontechnical people are spending a great deal of time and money to join these communities.

As we have seen in virtually every other medium, change is not neutral. It always has consequences. As America logs on and plugs in, will it improve our nation, or will damage the beleaguered thread of American identity? Since this is such a widespread phenomenon, we need to know what will happen to our notion of "community" as the nation goes digital.

Over the past few years, I have had the privilege of witnessing the birth of such a community. In the spring of 1994, I was hired by a local television station as a summer intern to create a prototype on-line service—a responsibility that kept me quite busy all summer long. As fall approached and I planned to return to school, I was asked to take the reigns of responsibility back with me to Washington, DC. After giving it my summer, I could hardly resist managing it for a few more months.

Over the next semester, I watched this technical project take on a life of its own. I saw other people invest their time, their energy and most importantly, their reputations into this system. In short, I witnessed the birth of a community.

To devote more time to the project, and the resulting community, I dropped out of school, established a consulting firm, and moved back to Minneapolis. Participants on this system knew that I had moved, but everything remained as it was. The community remained intact.

After 9 months, 1,000 registered users, and $100,000, it became quite evident what the kind of system this television station wanted to establish. It was equally evident that those in charge would not be able to bear the responsibility to manage, update, and maintain it themselves. Fortunately, another consulting group came to the rescue with the funds, expertise, and staff to bring this project to the logical next step in its development. In the process, I was fortunate enough to drop my titles of "moderator" and "nerd" to focus solely on content. It became my job to get the best content on our site, and help my coworkers enhance their own content.

The first thing our newly formed group decided was to abandon the technology of bulletin boards for the World Wide Web (WWW). It was a fairly simple decision, but by doing so, we sacrificed some well-developed tools for community building; the technology of the WWW was not advanced enough to facilitate chats and threaded discussions very well. However, the WWW had the potential of bringing our service to a much broader audience—the entire world.

Migrating onto the WWW, we decided to shut down our bulletin board system after 1 year of operation. But in that time frame, I had an incredible vantage point to witness several hundred people explore the new technology, interact with each other, and define their own notions of community.

Through my experiences with that group of people, I came to define virtual communities as a group of people connected by linked computers. These people may be in a geographically similar area. They may know each other. They may even keep in contact outside the computer network. But for the purposes of this chapter, any communication between humans over a computer network falls under the category of a virtual community.

In defining a community, we must refine it to its most basic elements: A place where people interact with each other. This can be in a bar, at a church, or at an amusement park. Whether these events occur in physical locations or through on-line environments, people will make some sort of contact with those around them; on bulletin boards, on Internet usenet groups, and in Multi User Dimension groups, people will react to the presence of others, and everyone's experience will be altered. Strange as it may seem, that is a community.

But perhaps an even more challenging task is to define the "real world." Virtual reality may closely approximate that of physical reality; thus, the definition of reality cannot rely on physical touch or appearance. Furthermore, the definition cannot relate to the use of intermediary equipment, as people in the "real world" often deal with each other through the mail, over telephones, or through an intercom. "The real world" seems to be defined by the absence of computer-aided mediation or translation. In "the real world," computers can neither translate communication nor alter it.

When people converse on a telephone (and both people represent themselves honestly) the only limiting factor is the extent to which their voices are distorted by the quality of the sound. In letter writing, correspondents can still represent themselves through extra-textual channels such as letter size and shape, style of handwriting, and the type of paper used. However, it seems unlikely that many would refer to such communication activities in terms of creating a community. Text-based systems are perhaps the biggest distorter of reality, and thus farthest away from "the real world." Not only does the print medium reduce contextual expression, it also allows users to distort their own identity. It is much easier to misrepresent gender, ethnicity, and appearance through text-based systems than it is through other modes of communication. Thus, by this definition, text-based interaction distinguishes itself from "the real world" when communication is altered by computers.

Rather than go to great lengths to further define what a virtual community is, it may make more sense to see what it does. The first likely result of increased participation in virtual communities is a blurring of identity, however identity is defined. As more people get involved, it will be increasingly usual to speak of "friends" whom you have never met. This can easily slip into the vernacular; at first it goes unnoticed. But gradually, people speak of friends or discussions with other people—without ever meeting them in person. And rather than focus on appearances, their interaction depends on the information transmitted.

In virtual worlds, people speak of "reading something" or "talking to someone," but on further investigation, these encounters are purely vicarious. The word "read" refers to transitory digits on a screen. And the "conversation" doesn't involve a physical meeting. But a community is established nevertheless.

In my own experience, I have noticed this phenomenon repeatedly with the on-line system I helped create for the television station. People referred to "friends" on line and said they "missed people" if users didn't participate very often. Many participants had regularly scheduled "chat times" when they offered to "talk" to others, and if people failed to participate, their presence was distinctly missed.

Feelings can be hurt, as well. On the system I moderated, people went to extreme lengths to defend their on-line personas. Just a small offhand remark could spark a cantankerous debate over one's reputation. This may seem a trivial matter—the offending and defending of on-line reputations. However, in those environments, characteristics such as wealth, appearance or strength are largely ignored. Instead, the most important quality to have in on-line environments is credibility. An assault on that is an assault on one's entire on-line existence.

The on-line service I helped create was originally envisioned as a system for news retrieval. But its essential fabric changed, as users revealed themselves in the chat areas, expressed opinions in the discussion groups, and placed emotional stock in their dealings on line. In fact, in the year that this system existed, users visited both ends of the emotional spectrum; for some, this computer network seemed to give them stability and hope. For others, it led to hurt feelings and shattered expectations.

And the most amazing aspect is the lack of technically oriented people. It would be easy to dismiss computer-mediated communities as solely for "nerds" or "techies." But this system was designed specifically for nontechnical people.

Perhaps more interesting is the way that these nontechnical people have experienced the same phenomena as have the "techies." Regardless of the technical acumen of the participants, the same social phenomena occur. One can only imagine the degree to which the United States will participate in national versions of such systems, and ultimately, be affected.

It soon becomes quite obvious how these situations will alter radically our notion of community. People are increasingly able to have "good friends" from around the world—as easily as from the opposite side of town. Distance is being removed as a barrier to communities. Instead of relying on geographic similarities, communities are now formed upon shared interests and experiences, or simply the perceptions thereof. In my experience, people prefer shared emotional or intellectual interests over geographical similarities with astounding regularity. I have seen people from disparate locations join in heated discussions—the same issues they may feel uncomfortable discussing with their neighbors.

The second result of virtual communities will likely be a widespread crisis of identity. When people enter these digital worlds, each person has an opportunity to define his or her own identity. Often, braggadocio emerges. Most identities "are the products of the players' own imagination and usually indicate the possession of attractive and even superhuman attributes" (Reid, 1995, p. 178).

This may seem like childish boasting, but it is a fairly common human reaction. For less technical displays of the same behavior, one need only look at telephone dating services or even singles' advertisements in the newspaper. There, people commonly exaggerate, or even lie, to make themselves seem more attractive to others. The main difference between those environments and virtual ones, is that users of dating services will, assumedly, eventually encounter one another in person. The same does not hold true for virtual communities. On-line groups are often quite content to meet solely in cyberspace. And without some degree of "real world interaction," one can only imagine the extent to which virtual boasting will surpass its telephone or newspaper contemporaries.

Reid (1995) suggests that this behavior is necessitated by the lack of other attributes. As no one else can determine body type, hair color, or tone of voice, users must funnel all of their attributes into one medium. And in that way, computer mediated communication is one-dimensional (p. 179).

As people spend more time on line, and develop a sense of themselves in their interactions, they may struggle with a conflict of identity. For someone to introduce herself (or himself) as "a tall, luscious young woman with long, wavy deepauburn hair that gleams golden when it catches the light..."(Carlstrom, 1992, p. 15) it may be difficult to return to being "a homely nerd" (as the same person later identifies herself). It may lead to a heightened awareness of appearance, as people attempt to live their virtual personas. Or, it may cause people to place all of their self-worth in their on-line identities, thus ignoring their own body. It will be interesting to see which side effect is most common.

In the same way that conceptions of beauty are irrevocably altered in virtual communities, so are notions of wealth and power. As information becomes more powerful, so do people with information. Authorities on technology, or "wizards" as they are frequently called, could garner a great deal more respect than an average millionaire. After acquiring the necessary equipment to join these worlds, money generally becomes irrelevant. It cannot purchase respect or obedience as easily as in traditional society.

Richard MacKinnon, an astute observer of virtual communities, agrees: "It is not likely that external world nobility will have relevance to [online communities]....Perhaps this is because persons of nobility, accustomed to the "trappings" of the elite, find that without these trappings, their nobility is nothing more than words" (p. 124).

Another outgrowth of virtual communities is the heightening of material world emotions. These technologies may heighten a feeling of euphoria, or might contribute to a sense of aloneness similar to a long-distance relationship. If one hangs up the telephone after a long conversation with a good friend, the amicable aura tends to linger. If, however, the conversation becomes antagonistic, or—even worse—there is no answer, the sensation of separation is heightened.

After an enjoyable stint on line, it can be very relaxing to savor the previous few hours, all in the privacy of an off-line home. On the other hand, a bad experience, a virtual harassment, or a virulent "flame" can seem more harmful than a verbal reprimand in person. It is irrelevant whether or not these emotions come from "the real world." Emotions are still determined by our perceptions of a situation. And in dealing with on-line environments, emotions are just as real—even when the community is virtual.

Using the hackneyed analogy between virtual communities and highways, we observe the same result: "While promising to bring us closer, highways in fact cater to our feeling of separateness" (Patton, 1986, p. 20). The same can be said about virtual communities. As promised, they can produce a feeling of intimacy. At the same time, however, they can strip it away and heighten the sensation of loss.

WORKS CONSULTED

Baym, N. K. (1995). The emergence of community in computer mediated communication. In S.G. Jones (Ed.), *Cybersociety: Computer-mediated communication and community* (pp. 164–183). Thousand Oaks, CA: Sage.

Carlstrom, E. (1992). The communicative implications of a text-only virtual environment. (Electronic manuscript.) Available via anonymous ftp to parcftp.xe-
rox.com from directory pub/MOO/papers.

Crichton, M. (1993). *Disclosure.* New York: Bantam.

Crichton, M. (1990). *Jurassic Park.* New York: Ballantine.

Dizard, W. (1994). *Old media/new media: Mass communications in the information age.* White Plains, NY: Longman.

Gibson, W. (1984). *Neuromancer.* New York: The Berkeley Publishing Group.

Huxley, A. (1934). *Brave New World.*

Jones, S. G. (1995) *Cybersociety: Computer-mediated communication and community.* Thousand Oaks, CA: Sage.

Jones, S. G. (1995). Understanding Community in the Information age. In S.G.
Jones (Ed.), *Cybersociety: Computer-mediated communication and community* (pp. 10–35). Thousand Oaks, CA: Sage.

Lehman, B. A. (1994). *Green paper: Intellectual property and the National Information Infrastructure.* Washington, DC: Information Infrastructure Task
Force.

MacKinnon, R. C. (1995). Searching for the Leviathan in Usenet. In S.G. Jones
(Ed.) (1995), *Cybersociety: Computer-mediated communication and community* (p. 112–137). Thousand Oaks, CA: Sage.

McLaughlin, M. L., Osborne, K. K. & Smith, C. B. (1995). Standards of conduct
on Usenet. In S.G. Jones (Ed.), *Cybersociety: Computer-mediated communication and community* (p. 90–111). Thousand Oaks, CA: Sage.

Neuman, W. R. (1991). *The future of the mass audience.* New York: Cambridge.

Patton, P. (1986). *Open road.* New York: Simon and Schuster.

Reid, E. (1995). Virtual worlds: Culture and imagination. In S.G. Jones (Ed.), *Cybersociety: Computer-mediated communication and community* (pp. 164–
183). Thousand Oaks, CA: Sage.

Toffler, A. (1990). *Powershift.* New York: Bantam.

6

Going On Line with the U.S. Constitution: Gender Divisions in the Cultural Context of the First Amendment

Kerric Harvey
The George Washington University

THE PROBLEM OF COMPETING CONSTITUTIONAL RIGHTS

Much of the intrinsic tension between those who defend U.S. free speech rights and those who are concerned with protections against hate speech and other harmful utterance revolves around competing interpretations of the First Amendment. These different ways of understanding the constitutional guarantees of free expression hinge, in turn, on conflicts between absolutist and relativist views of the Constitution—whether the letter or the spirit of the law should be the guiding criteria for judicial action.

The issue reduces to a difference of opinion about how to "read" constitutional texts, particularly as the United States continues to diversify and as the social contexts within which its governing documents are situated become ever more complex. Additionally, the evolution of an avalanche of new communication technologies, loosely grouped under the heading "on-line media," compel media practitioners and policymakers alike to revisit several of the fundamental underpinnings of traditional free speech concepts.

The issue of how to conceptualize the Constitution—and by extension, the First Amendment—has always become problematized by the societal attitudes toward women and minorities. The historically consistent underprivileging of women, in particular, suggests how very urgent it might be to re-frame the First Amendment in the light of a commercially based press and an increasingly technologized na-

Note: Some of the ideas explored in this chapter were first presented in the Law Division of the Association for Education in Journalism and Mass Communication, 1995 annual conference, in Washington, DC, August 8–14.

tional media system. Even purportedly "democratic" media, such as the Internet and the World Wide Web (WWW), are complexified when their communicative advantages are applied to historically disenfranchised groups. Examining the specifics of how this plays out in the case of women highlights the key issues and persistent concerns of many marginalized groups, who often find themselves juggling the "pros" and "cons" of on-line journalism and Internet-based news delivery.

WOMEN AND THE U.S. MEDIA SYSTEM

Excluded sometimes by law and often by custom from the normative press of public affairs, women have, historically, been pushed to the fringes of American social and political debate, suing for a more central role and, when that role was repeatedly denied them, creating alternative arenas of their own. In law, in education, in medicine, in social issues and in religion, in the workforce and, when possible, in politics, women built decision-making and organizational structures of their own, dealing with their exclusion by making it irrelevant to their goals (Borden, 1993).

In many ways, the female world worked tremendously well, and important vestiges of it linger today in the networking resources and the public interest institutions enjoyed by the social heirs of these early agitators. But at least one barrier to real power has endured throughout both cycles of the so-called "women's movement," persisting into the contemporary political scene and extending to the online arena. Ironically, this obstacle to full female enfranchisement is the same governmental guarantee that promises all U.S. citizens an equal voice in the affairs of their society—the First Amendment to the Constitution of the United States.

Conventional academic wisdom (see, e.g., Fishman, 1980; Gitlin, 1980; Tuchman, 1978, among others) maintains that the remedy for biased or inadequate media coverage is found in increasing the true diversity—not just the multiplicity—of voices participating in the national conversation, suggesting that equal access to media translates to equal impact on public opinion. This is an important point to explore more thoroughly.

Increasing opportunities for media access is certainly a critical component in righting the persistently imbalanced "playing field" which has, historically, characterized so much of the American media scene. It is equally important, however, to acknowledge that the access/impact equation is not, perhaps, quite as straightforward a calculation as it may at first appear. This is because there is a exponential factor in the economic and political formulae from which the media/impact equation is derived. This exponential variable has been a hidden but critical factor in calculating the impact of media messages in the traditional media environment—print, broadcast, and cable—and is all the more vital an ingredient in the approaching on-line era.

At the heart of this equation is the idea that the composition of any given media audience matters, in the political sense, just as much as does its size—all slices of the media audience, in others words, are not created politically equal. This is an odious but inescapable aspect of democratic monopoly capitalism. Reaching 20,000 socially and economically marginalized people with one's political message may, for example, have less real impact than reaching 200 key national decison makers. And here is where a closer look at the gender politics of media access becomes an absolutely vital part of any discussion about First Amendment equity.

ON-LINE TECHNOLOGIES AND THE AMERICAN POLITICAL ELITE

The unfortunately limited database available regarding exactly who uses on-line media underscores the importance of recognizing this functionally important power differential between theoretically equal elements in the U.S. electorate. Gender, clearly, is a key demographic illustrating serious divisions within the supposedly intact body politic; far more men than women use the Internet on a regular basis. About 18% of America's men report using a computer at home "almost every day," as compared to 9% of U.S. women. Additionally, more people of European descent than any other racial or ethnic group are home computer owners and habitual Internet users, a trend which appears to extend to the next generation of Americans. About 35% of White households have a child using a computer, as opposed to 18% of Black households. When probing the numbers for slightly older children, 30% of Black teenagers, as compared to more than 50% of White teenagers, report access to a home computer (Times-Mirror, 1994, p. 30).

Socioeconomic status (often related to both gender and race is a key factor in one's chances of inclusion in the elite on-line community. A detailed research project by the Times-Mirror Center for the People and the Press (1994) found that, "Only 11% of the least affluent households (income under $20,000) has a PC, and an identical 11% of homes in which the respondent had not finished high school has a PC. In contrast, over half (56%) of households with a family income above $50,000 has one, as do almost two-thirds (65%) of respondents with at least some post-graduate training" (p. 39).

Not only do conspicuously more college graduates own the home computing machinery that is vital for on-line connectivity than those who did not go to college, the 1994 Times-Mirror project found that: "...the spread of technology through American society is quite uneven....A college graduate with a family income of $50,000 a year is three times more likely...than a nongraduate who earns less than $30,000...to own (all types of electronic technology)....The gap is nearly five to one for personal computer ownership and an enormous 10 to 1 for on-line capability within the home" (p. 8).

This disparity is even more pronounced for modem users—the people who have a demonstrated capability to hook their home computer into the on-line arena. Most modem users are male (69%), more affluent than the average person (53% make $50,000 or more a year, compared to the national average of 23%), and younger (about 35% of the general public is over 50 years old, but only 17% of modem users are). Only 5% of modem users are Black, as compared to about 13% of the general population; about 28% of the nation's Hispanic population reports having a home computer, but only 8% a modem, in the home (Times-Mirror, 1994, pp. 34, 47–48).

Age and geography are other important lane markers for traffic flow in the American on-line environment, what the press and public alike have come to call the "Information Superhighway." Outside of the Atlanta technology corridor, people over 65 and people living in the American south are chronically underserved by on-line infrastructures. Computer ownership among those 65 or older hovers around 10%; about 27% of Southern households report home personal computer (PC) ownership, as compared with 35% in the east (Times-Mirror, 1994, p. 47).

Why should this matter? It gets back to the question of encouraging truly democratic dialogue in a self-governing nation. At the time of this writing, the accepted figure for U.S. Internet connectivity hovers around 24 million people in U.S. and Canadian households combined (Kim, 1995, p. E1), with home PC ownership saturation registering at about 35% in the United States. Overall, Internet connectivity is around 10% for U.S. sites (Kornbluth, 1995, p. 36), with some sources claiming that even by the year 2000, only 22% of all U.S. households will be linked to the Internet in a usuable way (Kim, 1995, p. E2). And even if the saturation rate is higher than that, the question of just which households are on-line, and which are not, remains culturally salient and politically troubling. Of the 30 million or so people who have access to the Internet, and the 11 million who live in the United States (Kornbluth, 1995, p. 40), only 2 million can afford the hardware and the connection time needed to participate in political debates and activities, such as national elections, which include an on-line component (Powers, 1994, p. E6). That's less than 1% of the total U.S. population, and, furthermore, this 2 million is comprised of those who are in a financial position to already have at their disposal multiple sources of political information and access to the social elite who create that kind of material.

The bottom line is the realization that the "we" to whom Internet aficionados constantly refer actually represent a profoundly skewed segment of the American population, which, perhaps sincerely, mistakes itself as being representative of "all" Americans. Even assuming all best intentions, this narrow slice of the demographic pie is busily engaged in making on-line policy on the basis of an illusion—policy which cannot help but aggravate further the dramatically different experience of cyberspace between White, male, college-educated, and financially secure cyber-citizens and all those many others who reside in a different demographic neighborhood.

Since, aside from pricing debates, the most robust controversy in on-line conversation seems to center around pornography and free speech concerns, it's important for policymakers to grasp fully the historical context of the enduring—and highly unfortunate—free speech versus safe society debate. In many ways, this controversy provides the classic example of the effect of positionality on political perception—the idea that who we are affects the way we see the issues. As such, it suggests some key concerns for future policy in other areas of on-line media enterprise.

WOMEN AND THE FIRST AMENDMENT—THEN AND NOW

In its most fundamental form, the issue reduces to a struggle between what are, essentially, competing human rights. Borden (1993), Byerly (1993), and others have done work which suggests that one of the enduring drawbacks to the exercise of free speech rights in the postindustrial United States is that it tends to take place within a larger context of environmental misogyny.

This unfocused malevolent atmosphere breeds an unhealthy antagonism between the right of women to be treated as social equals and the "right" of some men (and perhaps some women) to say harmful things about women as a group—and to own, literally as well as figuratively—the media channels through which those hurtful messages are to distributed. For despite the best intentions of the Federal Communications Commission, the Commerce Department, the Securities and Exchange Commission, and the other regulatory bodies charged with the responsibility for spreading around the opportunity to own a piece of America's media, actual media ownership has consolidated at an alarming rate.

Bagdikian notes (1987) that about 29 people now control most of the Western commercial media—and he expects that within less than 20 years, that number will shrink to fewer than 6 individuals controlling the majority of information and media outlets. Of those figures, industry numbers suggest that women still trail in the ranks of either media owners or media top management (Dickson, 1993).

And as other scholars (Bennett 1983; Gitlin 1980; Lang & Lang, 1983; Tuchman, 1978) have pointed out, ownership per se is not the only constraint on marginalized groups seeking to be included in the national media dialogue. Often the routines of newsgathering itself converge with the professional culture of the journalism industry to squeeze out issues and individuals which challenge the social and political status quo. In this sense, the media are a sort of annexed territory to corporate American culture. And as both classical and new critical theory suggest, this is unlikely to change any time soon, since the professional routines of commercially based, deadline driven journalism tend to reward—with praise, promotions, choice assignments, and sometimes, outright cash prizes—those journalists who conform most precisely to the media's view of the social world. These and other factors converge to create an essentially institutional culture within which media work is actually conducted. Not surprisingly, this yields media organizations characterized by a group worldview that is, by and large, consistent with the worldview of other large, bureaucratic institutions of social power (Gitlin, 1980, Tuchman, 1976).

Deciding what is and is not "newsworthy," deciding who is and is not a "credible source," forging the narrative reconstruction of "newsworthy" events, and other day-to-day details of doing journalism impose enormous restrictions on the media's already limited vision, since by institutional definition, "credible" sources tend to be industry leaders, public officials, experts, organizational representatives (but only of officially "credible" organizations), and involved parties. And "newsworthy events" tend to be those of importance to these types of people. That leaves most of the public—and most of the world—out of the journalistic loop.

This supports the notion that systematic distortions of First Amendment principles—equal and meaningful access to and participation in the process of free expression and significant public debate—already occur in at least two places, at the level of media ownership and within the daily routine of media practice. Further complicating the picture is the growing presence of certain types of expression which are, either by definition or in a collateral sense, intended to express the speaker's intention to harm someone else, rather than to contribute to constructive social discourse.

Hate speech certainly falls into this category, since the express purpose of such utterance is solely to demean or diminish or to damage outright the members of a group someone else finds objectionable. At the time of this writing, however, civil liberties groups such as the Electronic Frontier Foundation (EFF) and the Progress and Freedom Foundation, are fighting to achieve the intact and immediate importation of free speech protections from preceding technolgies into the cyberspatial realm. The Progress and Freedom Foundation has gone so far as to develop a "Magna Carta for the Knowledge Age" (1994), containing essays by the famous futurists Alvin and Heidi Toffler. Normative power is ascribed to on-line information technology in that document:

> The central event of the 20th century is the overthrow of matter. In technology, economics, and the politics of nations, wealth—in the form of physical resources—has been losing value and significance....As humankind explores this new "elecronic frontier" of knowledge, it must confront again the most profound questions of how to organize itself for the common good. The meaning of freedom, structures of self-government, definitions of property, nature of competition, conditions for cooperation, sense of community and nature of progress will be redefined for the Knowledge Age—just as they were redefined for a new age of industry some 250 years ago (p. 1).

According to on-line absolutists, in the 21st century information will be sovereign, and communications technology the royal road linking those who govern with those who are governed.

There has emerged, however, an unexpected challenge to technology's uncontested dominance. In the mid-1990s, growing concern for protecting America's children from the damaging effects of sexual and/or violent media content, and from being cast as attractive targets for sexually inappropriate adults, reached a point of critical mass, culminating in a series of Congressional hearings on the potentially harmful effects of a media system drenched in sexual and/or violent imagery. This provided a challenge from an unexpected quarter—the nursery—to the absolute sovereignty of informationalized culture.

Both the direct impact of media content on the personality development (in this case, of the child) and the residual effects of media imagery on the social context within which the personality is situated (the larger society) were perceived as places where safety concerns appropriately overrode First Amendment prerogatives. Evidence for this shift in Congressional thinking can be found in Section 551(a) of Title V, embedded within the Telecommunications Act of 1996.

Interestingly, much of the scholarly, ethical, and philosophical terrain explored within Congress' investigation of the children's television debate replicated the ground covered—and subsequently abandoned—when the same questions were raised regarding potentially deleterious effects of violent pornography on women, rather than on children. In terms of the social science involved, there is little—if any—substantive difference between what can and cannot be "proved" as a conclusive social outcome of harmful image representation for children as opposed to for women. Why different thresholds of "prove-ability" are required for the two different at-risk groups remains a mystery, the answer to which suggests a whole range of uncomfortable and fundamentally misogynist cultural dynamics. Although it is arguable that children as a group may be more socially vulnerable than women in American society, crime statistics against women, specifically, suggest that this is a surface detail only. Be that as it may, however, with the passage of the 1996 Telecommunications Act, Congress has now acknowledged that physical-world predatory action toward children, at least, may be catalyzed by violent media imagery.

How might media-catalyzed harm translate to the on-line realm? Several options present themselves. Before a particular regulatory avenue is selected, however, decisions must be made regarding the essential nature of cyberspace itself. At the top of the list is the most basic question of all: Is cyberspace a "thing," or is it a "place"?

If the answer comes down on the side of "thing"—if cyberspace is deemed a communications medium—then the next round of distinctions must decide what type of antecedent media on-line technologies most closely resemble: print, broadcast, or common carrier-type technologies? In that case, the regulatory options congregate around familiar, although thorny, free speech principles. Libel, slander, defamation, clear and present danger/fighting words restraint, truth in advertising, intellectual property, privacy concerns, and obscenity, pornography, and decency restrictions are invoked as some of the instances in which prohibitions against speech are constitutionally justified.

One more aspect of the pornography per se discussion must be raised here, because it points up the astonishing potential of emerging communications technology to transform even the most fundamental concepts employed in baseline free speech issues. A compelling argument against extending constitutional protection to child pornography has been the physical welfare of the child in the creation of such media products.

Even those who remain unmoved at "snuff films," which feature women being, literally, murdered onscreen during sexual intercourse, may balk at the notion of a child being coerced into sexual activity for the benefit of the camera. But modern digital technology introduces the troubling possibility of producing child pornography in which no real children are physically present. What might this do to the one area of persistent restriction on sexual representation? Removing the physical child from child pornography may protect one individual boy or girl, but it doesn't do much for altering the overall media environment in which the concept of coerced sex with boys and girls—or with adult men and women—is reinforced regularly.

Still unaddressed, however, is the question of whether cyberspace is a means or an end, a place or a medium. The question is even more salient, perhaps, as phenomena like digitalized imagery and distributed virtual reality continue to press the boundaries between utterance and experience, both of which are increasingly deliverable through on-line means.

If the answer is that cyberspace is more like a physical place, then an entirely new set of criteria must be developed for assessing the relative rights and responsibilities of those who do business, conduct relationships, or otherwise go visiting there. Stalking laws, trespass restrictions, remedies for assault and battery, and other types of behavioral injunctions based on bodily safety and physical violations would then supply the prevailing regulatory guidelines.

POLICYMAKING BY DEFINITIONAL DEFAULT

An often overlooked, but legally critical point in framing cyberspace as a medium, not as a place, lies in the operational irrelevance of First Amendment law to most of its denizens. Accusations abound that systems operators (and other agents of the on-line provider companies) violate free speech rights when they moderate cyberspatial conversation, as in cases of deleting abusive or sexually explicit material on Bulletin Board Services. But the hard fact is that the First Amendment applies only to governmental actions, not to the behaviors of private companies or individual citizens (Cavazos & Morin, 1994, pp. 69–70).

Prior to the Telecommunications Act of 1996, the First Amendment was widely understood as applicable only to restrictions on speech undertaken by government representatives; it had no authority over private sector media organizations. An especially ill-thought out section of the 1996 Act, however, effectively shatters this one true thing about U.S. communications law. Title V of the 1996 law, subtitled the "Communications Decency Act of 1995," provides explicit injunctions against person or persons who knowingly use a telecommunications device or system to conduct a variety of irritating or malevolent things, including but not limited to those who: "...make, create, solicit, and initiate transmission of any comment, request, suggestion, proposal, image, or other communication which is obscene, lewd, lascivious, filthy, or indecent, with intent to annoy, abuse, threaten, or harass another person..." (p. 81).

The Act makes added and specific provisions against executing such prohibited behavior against minors, especially, and further specifies that these restraints on speech occur when a person "...makes a telephone call or utilizes a telecommunications device, whether or not conversation or communication ensues..." (p. 81). Liability of both the carrier organization—the on-line provider—and the offending individual is spelled out.

What the lawmakers have either missed or have chosen to ignore, here, is that the providers are, overwhelmingly, privately owned businesses. They were never at risk of First Amendment violation in the first place, since the First Amendment applies only to government regulation of speech and private carriers do not fit into that definitional category.

What does this mean? Several things. First of all, it means that it will hurt less when the Act is overturned by "expedited review"—a revisiting of its constitutionality that is expressly provided for in Subtitle C, Section 561 of the overall act. Under this process, the usual lengthy and cumbersome method of examining complicated questions like constitutionality is "streamlined" into a dramatically truncated version of the usual procedure. Subtitle C is very short, as is the process it guarantees to deliver. It reads: "Any constitutional challenge to any part of Title V is to be heard by a three-judge District Court panel, and will have a direct appeal to the Supreme Court" (p. 91). That's subsection (a). Subsection (b) states that such appeals to the Supreme Court must occur within 20 days after the entering of the contested judgment at the lower court level.

In essence, this provision constructs a breathtakingly elegant regulatory loophole, through which members of an election-year Congress can march righteously on their way to home state stumping grounds. Subtitle C transfers the political hot potato of the Decency Act from the Congress to the Supreme Court, which does not face re-election pressures in 1996, but which, like Congress, contains conservative elements itching for a watershed case through which binding high court precedent can be set.

Which of two core, and, in this case, competing conservative values will eventually prevail—exoneration of private sector social responsibility or the imposition of restrictive moral values on public speech—remains to be seen. One likely outcome from this elevated paper fray, however, is the likelihood of the Supreme Court using its judicial authority to rewrite that part of the Decency Act (Sections 502 and 223) which holds the provider—in this case, business interests—liable for obscene, indecent, or pornographic material passing through its on-line channels. In other words, practically speaking, Supreme Court intervention completes the work that Congress did not dare; it supplies an occasion to award telephone and cable companies, and other on-line providers, with broadcast media-like financial opportunities, while at the same time investing them with common carrier-like protections against responsibility for transmitted materials.

Two other, more long-term outcomes may be projected from Congress's application of First Amendment issues to private industry entities. The first of these is the implication that the conceptual—and cultural—distinctions between government-owned and privately owned enterprises are becoming less and less relevant. Since Section 254 of the Act all but sanctions corporations like AT&T as a privatized arm of the federal government, by making universal service an official national mandate, it's arguable that this type of public/private sector conflation is indeed an intentional act on the part of the 104th Congress.

The second long-term issue returns the discussion to trying to apply First Amendment principles to private corporations. The tactical implications of this enterprise have already been explored. Strategically, however, it may be equally important for corporate America to force the issue of nonculpability in this most public and most conclusive of all arenas—that is to say, the U.S. Supreme Court.

Since, by definition, attempts to subject private industry to government standards of nonintervention in the exercise of free speech are doomed, it is precisely by going through the motions of doing so that private industry may become permanently freed from all responsibility for any social ill effects of that "hands off" policy. This would finally "settle," once and for all, the naggingly complicated question of just who is responsible for protecting against harmful speech, by setting up an indefensible position which, when knocked down, yields very real and utterly airtight law, since it was "decided" in the nation's highest court.

All this is applicable, of course, only if cyberspace and the utterances contained therein are defined as communication channels, not as social places.

CYBERSPACE OR CYBERPLACE?

If cyberspace is defined as an actual place, rather than a mass medium, of course, an entirely different approach must be developed to guide the conduct of common life within its bounds. Unfortunately, the signs are not good that the full complexity of this challenge is realized by those who seek to address it. Threat in the physical realm carries a graver penalty than that sustained through mediated channels. In that regard, it is not off-target to consider hate speech, violent pornography, and electronic voyeurism as naturally evolving versions of their physical counterparts.

Stalking has always had more female than male victims, so it's only logical to assume that men and women will react to the threat of stalking in somewhat different ways. On-line stalking is no exception to this rule. But the persistently blurry conceptualization of cyberspace makes seeking safety from on-line stalkers even more difficult than finding relief from material-world predators. Laurie Powell's ordeal (Mark, 1995) in search of protection from an on-line assailant illustrates this point with horrific clarity.

In February of 1993, Laurie Powell, a Greensboro, North Carolina housewife, was browsing through the usual array of user groups and bulletin boards. She was shocked to find herself named as the subject of an angry and obscene message posted on the Home Life and Medical bulletin board, where she often sought conversation on resources and strategies for raising her retarded son.

Too upset to share the event, she told no one about it. But more abusive postings appeared, one of which included Powell's home address. When she contacted Prodigy, the company informed her that the sender had used a fake ID, precluding real-world identification. The company kicked the message off-line after about 3 hours—plenty of time for it to be seen by any of Prodigy's more than 2 million users.

But on-line assault was not the end of the story for Powell. Before long, it became clear that "Vito"—as the increasingly vicious and increasingly specific messages were signed—was putting together material on Powell's off-line life, as well. In addition to making sexual threats against her 12-year-old retarded son and teenage daughter, Vito and other, equally ephemeral message-senders began tinkering with Powell's credit rating. When she was rushed to the hospital for emergency surgery, an event about which she told no one except her family, her hospital room and telephone number appeared on the Net within 24 hours, indicating that her medical files were being constantly cross-referenced against local medical facility admitting data.

Other women became targeted by Vito and "Vito-wannabes," sometimes with information as specific as the mileage rates of their current automobiles. An attempt was made to lure Powell's daughter, Amber, away from her high school in the middle of the day; in a related nightmare event, the outside screen on Amber's bedroom window kept disappearing.

Powell kept pressing for help. Prodigy kept telling Powell that it had taken "appropriate action," saying it had yanked the various accounts under which Vito messages kept appearing. But still the harassment continued. Complaints from a now terrorized Powell, and other women being subjected to "Vito" and "his" brethren, prompted the company to launch an internal investigation, eventually bringing in a Fresno Police Department computer crimes expert, Frank Clark, to pursue the matter further. On the grounds that it performed a common carrier function similar to the telephone company, Prodigy declared itself free from liability for the content of any messages carried through its services.

Frank Clark, himself, wound up being pursued, logging on to his personal e-mail account one day to find sexually explicit graphic images. Undaunted, Clark sorted through pounds of computer printouts and unsnarled a big chunk of the thousands of daily log-ins snaking through Prodigy's service in 1993. Eventually he narrowed his short list of suspects to Mark Johnson, a Mormon substitute teacher living in Fresno.

Johnson admitted to using the name "Vito" on line, but denied authorship of the avalanche of abusive messages that had been plaguing the Prodigy system. Clark was unconvinced of Johnson's innocence, and continued exploring the possibility that Johnson was at the root of Powell's harassment. Eventually Clark accumulated enough testimony and support material to get a search warrant for Johnson's apartment.

A search of the place revealed enough high-tech equipment and questionable data records to convince Clark that Johnson had perpetrated credit card fraud and electronically based grand theft. But a week later the Fresno district attorney disagreed and tossed the case out on grounds of insufficient evidence—or, as Clark suggested, too complicated, too technological evidence, around which there exists no clearly defined body of supporting law. No one really wants to handle crime in cyberspace—not local police departments, not the FBI, not Congress, if the Telecommunications Act of 1996 is any indication—because they are sticky affairs that occur in boggy legal terrain.

Laurie Powell's gruesome story, taken as a parable about just how distressingly real a "virtual" on-line experience can become, returns the discussion to potential deficiencies in the way in which U.S. law has understood the cultural dimension of the free speech/pornography relationship, per se.

PORNOGRAPHY AS HARMFUL SPEECH IN ON-LINE ARENAS

The traditional focus of the free speech/fair society debate has been pornography, particularly in its more violent and misogynist incarnations. Ironically, it is within this most private arena of human experience—the bedroom—that this most public of social debates is persistently pursued. Adding sex to the already volatile blend of competing human rights does two things simultaneously.

First, it further muddies waters that are fairly murky to begin with, because it collapses into the free speech issue an entire range of sweeping cultural attitudes—and the problems which attend them—regarding the fundamental imbalance of power between genders, reflected many places. Granted, it's important to recognize that overarching social inequities extend all the way through the social structure, invading private lives as well as defining public ones. But in a sense, this is like trying to "fix" what's on TV by re-programming one's VCR to record at a different hour—the baseline problem remains the same, no matter at which point it intersects one's life. It may well be the case that violent pornography which victimizes women will neither persist because it is protected speech nor disappear if it is stripped of its First Amendment armor. Instead, it might only fade away if the cultural circumstances which make its consumption both personally satifying and culturally tolerable are eroded first.

Pornography will cease to exist not because it is legislated to death, but because its audience becomes either bored by it or ashamed when they consume it...when pornography isn't "manly," anymore, when it's reframed as an indication of insufficient manliness, of embarrassing sexual immaturity. And in order for that to happen, much larger cultural changes must occur first.

Aside from depending on this anthropological remedy to take effect, many studies have been done exploring the best way to protect the First Amendment's potency without marginalizing a significant percentage of the people who must live under its directive. By and large, these discussions focus on the changing nature of the American public sphere, citing profound shifts in both the body politic and the U.S. economic profile as reasons to revisit traditional interpretations of the Constitution.

But the second thing that happens when pornography becomes the fulcrum around which competing ideas of First Amendment interpretation are torqued is that a perhaps unnatural polarization emerges between the two primary constituencies of the American public—men and women. From the female perspective, the call has come out, repeatedly, for new and increasingly inclusive judicial rulings exploring the need for appropriate limits on speech in those areas where the exercise of free expression may cause harm, injury, or loss to institutionally disenfranchised subgroups within the U.S. workforce and the general population—and may cause harm to women and minorities, in particular.

But just as often as the question of reframing free speech rights has been raised, it has been buried. The deeply antagonistic nature of the relationship between First Amendment "absolutists," who protest what they term "censorship" in any and all forms and places, and those who place ahead of free speech rights a concern with personal safety and its ancillary social equities, has stalemated many an academic conversation and stymied many a well-intentioned public official. Precisely why this happens becomes somewhat more clear when the situation is examined not in terms of gender politics, but rather within the framework of gender logic.

Cate (1992) sums up the paradigmatic argument in favor of First Amendment absolutism when he addresses the reasoning behind protecting "free speech" which jeopardizes—psychically if not physically—the well-being of half the U.S. population. As Cate puts it:

> Consider the case of pornography....few defenders of pornography argue that the expression itself is so valuable that it should be permitted irrespective of its impact on society. But many commentators do not trust the government to draw lines around what is pornographic....It primarily reflects the concern that whenever we try to separate permitted from unpermitted speech, no matter where that line is drawn, every individual will at some point find some expression that he or she believes should be permitted....Thus, on the one side of the balance is the value of the pornographic speech plus the value of other speech that may be affected by the regulation of pornography. The other side of the balance contains the harm to society caused by allowing that pornographic expression. (p. 4)

Clearly, there is a glaring deductive flaw in this reasoning, which claims that in order for a law to be fair it has to make *everyone* happy *all* of the time. But Cate's statement also raises the issue of whether the problem of how to deal with pornography lies not in the definition of pornography itself, but rather in our understanding of "society." Gitlin (1980), Tuchman (1978), Bennett (1983), Newcomb and Hirsch (1987), and others suggest that the different sectors of society, which Cate and others think of as a homogeneous, monolithic bloc, may actually contain importantly different population subgroups. It's highly likely that each of these groups would entertain vastly different perspectives of what's at stake in the pornography issue.

As feminist scholars have long suggested, there is an unfortunate consistency among those persons who tend to be featured inaccurately and unflatteringly in pornographic representation. That common factor is gender. Women are more often and more likely than men to be portrayed in pornographic material occupying disempowered and victimized roles, which may be, to some extent, only the exaggerated versions of persistent institutional and societal victimization in other arenas. Given this, it is clear that women, as a group, are likely to react to pornography in a different way than are men, as a group, because: (a) the kinds of violation, intimidation, and intrusion that they see enacted against women in violent pornography are all too reminiscent of power imbalances they experience in their own lives, although perhaps in less physicalized form, and (b) men are not, by and large, on the receiving end of sexual violence in the kinds of "speech" one is dealing with in the pornography versus free speech debate.

But the social equation itself—the balancing act that Cate describes so succinctly—does not take into account these real and enduring differences in the context from which different parts of the society emerge as they grapple with the free speech/pornography issue. Liberal legal scholars have been wont to speak, somewhat blithely, of "society" as if it were a single, unified whole, but cultural theory suggests that this is not the case, and that different social groups may understand the risk/benefit ratio involved in protecting pornography in very different ways.

To many men and women, for example, the fundamental question regarding the relative benefits of permitting all speech for fear of not knowing what speech to prohibit may arrange itself in terms of contrasting priorities, depending on the degree of likelihood that they, themselves, might be the victim of harmful expression. For women, the sexual logic may read: "Free speech at any cost? Perhaps a better way to navigate the free speech/fair society paradox would be to give up some small measure of protected speech in exchange for an increase in public safety and personal dignity." This makes sense, since women are the ones whose public safety and personal dignity are more at risk from the effects of totally unchecked speech.

Men, far less likely to suffer direct personal consequences of pornography's virulent effects, may phrase the question exactly the other way around: "Fair society at any cost? The best way to achieve that is to encourage the free and unfettered flow of self expression, so that every voice in that society has an equal chance of being heard. It might get kind of loud and kind of rough-and-tumble, but...that's life." The key difference between these two attitudes, of course, is that what men might consider merely difficult or distasteful—speech which shows one being dismembered, or decapitated, or raped, or beaten—women are likely to consider outright dangerous to them, emotionally, socially, and psychically, if not in terms of providing inspiration for outright physical assault. Whether or not the legislation survives judicial review, in passing the Decency Act as part of the Telecommunications Act of 1996, Congress has slapped some backhand support onto this argument.

On-line media systems, in addition to supplying a massive new forum for the utterance and distribution of these messages, can also be useful tools for enacting violent fantasies in the physical world, where they cannot be undone at the touch of a button. Perhaps the most famous example of this chilling application of the Internet's ability to enter private lives can be found in the July 1989 murder of television actress Rebecca Schaeffer, who was shot to death outside her West Hollywood apartment by a man who had found her by hiring a private detective who specialized in Internet and computer-assisted investigations (Downs, 1996, pp. 35-36).

The overflow from on-line life to physical world is only one aspect of Internet regulation that needs to be addressed despite its rampant complexity. The whole idea of cyberspeech as a form of protected utterance—as an utterance, per se—needs to be revisited.

CYBERPORN AND THE PURSUIT OF FREE SPEECH

As already mentioned, on February 1, 1996, the U.S. Congress passed and President Bill Clinton signed into law the Telecommunications Act of 1996, which instituted sweeping changes in U.S. media law on every possible front, although the Act's precise effects are still to be unraveled. Although it may seem as if the bill's main contribution is to unbundle the tangled economic strands of the American media scene, the long-term effects, arguably, will reside in the cultural realm. That section of the legislation least likely to survive the implementation process is the Communications Decency Act of 1995, which would place transmission restrictions on objectionable speech sent through Internet channels.

It is the Internet connection itself that is of specific interest here. Earlier in this chapter, the point was made that many players on the American on-line scene claim some type of "public forum" status for the Internet, characterizing it as a "social place" rather than as a simple communications medium, and thus arguing that full First Amendment and right of assembly protections be awarded for all traffic flowing along its channels. The EFF, in particular, has waxed eloquent about the Internet's importance as a "...real marketplace of ideas, where the remedy for offensive speech is more speech" (Cavazos & Morin, 1994, p. 77). If this is, indeed, the case, then their claims that the Internet be exempt from restrictions on carrying speech based on pornographic, obscene, or indecent material may have some rhetorical weight. (For the purposes of main discussion, the complicated and legally important distinctions among pornography, obscenity, and indecent speech are herein collapsed.)

But if cyberspace is indeed a place more than it is a medium, as the civil libertarians insist, then debates about free speech rights on the Internet may be missing the mark altogether. Social debate regarding what can and cannot happen in cyberspace may be more appropriately housed under an entirely different umbrella concept.

If cyberspace is a place, then it is also a workplace, and the presence of violent or pornographic material in that workplace can thus be seen as a way in which the workplace itself is made hostile to women. This, in turn, means that permitting the presence of pornographic material on the Internet actually constitutes a form of sexual harassment.

Laws regarding sexual harassment in the workplace include, in their definition of what comprises an unacceptable level of transgression, specific guidelines for determining actions or words which create a "climate of harassment" for an identifiable segment of the workforce, such as women. How that principle of law could apply in the microcosm of the work environment, but not hold true in the larger universe of the nation which shelters those workforces, is a puzzle. And yet that is precisely what the emphasis on the "free speech at any cost, including pornography" argument insists on imposing—it provides license to create a national climate of harassment, within which individual instances of harassment are illegal. Laurie Powell's terrifying experiences on the Internet, discussed earlier in this chapter, illustrate clearly just how rabidly hostile to women the electronic public forum can become.

The problem with the sexual harassment argument against cyberporn, of course, is that sexual harassment regulation is a lower order of law than is the First Amendment—constitutional protections outgun statutory or administrative law, every time. But even given this, there still exists a whimsical kind of constitutional protection for harmful speech-targeted groups on a First Amendment protected on-line system, and the source of that fugitive protection is, oddly enough, the First Amendment.

A different part of the First Amendment. Free speech is not the only civil liberty guaranteed by that famous clause. The right to free assembly is also articulated, just as clearly, and with as much historical force, as is the right to free speech. If on-line harassment is profound enough to drive a particular subset of users, as a class, off the Internet, then a case could be made that the free assembly part of their First Amendment rights is being abridged by someone else's exercise of their free speech part of the same amendment. Now what? Suit and countersuit, it would seem...both bringing to bear the full weight of constitutional authority.

The spirit of much thinking in extant case law supports the idea that something must happen to provide some type of relief for the targets of group-oriented harmful speech, such as pornography. Cate (1992) notes that, to date, attempts to clarify the line between permitted and regulated speech usually rely on some set of "objective" standards that are supposed to distinguish between instances of speech that are dangerous and those that are merely distasteful. Like the courts before him, Cate characterizes this as a fairly impossible task, noting that:

> ...the (Supreme) Court frequently opts for no regulation of speech, even where the speech poses an immediate threat to society....What does this disappointing foray into First Amendment line-drawing reveal? One obvious lesson is that whatever their outcome, balances in the First Amendment are laden with value judgments. We may want to avoid such judgments, but *subjective assessments of competing interests are nonetheless unavoidable* (emphasis added). For instance, there is an assumption underlying a conclusion that no lines (drawn around speech) are better than imprecise lines. Such a conclusion suggests that we fear the dangers resulting from letting courts make subjective evaluations more than we fear the dangers resulting from pornography or other types of speech, irrespective of how violent or degrading to women it might be. (pp. 7–9).

The critical question to arise from this observation is: Who is the "we" to whom Cate refers in the last sentence? "We," historically, is men, who don't have to worry about the "dangers resulting from pornography" because—on line or off—they are not the subject of it, at least not in the same relentless fashion as are women. It may be that, to make the Constitution truly inclusive of all of its "wards," we must consider the possibility of—not rewriting—but actually adding to the original document.

MOVING TOWARD REVIEW:
HISTORICAL IMPULSES BEHIND CONSTITUTIONAL DESIGN

Tinkering with the Constitution by adding new clauses may not be as heinous or as heretical as it first appears. It's important to realize that the original document, itself, was the product of a long and tortured series of compromises between the newly independent colonies' knee-jerk reactions to more than half a century of burdensome British rule and the political ambitions and personal foibles of a few key men. A closer look at the historical context within which the Constitution first came to pass is an important first step in understanding that document's deeper cultural structure—a movement toward explicating its "hidden personality," as it were.

In their bull-headed insistence on controlling cross-Atlantic destinies from afar, the British unwittingly wrote the outline for the American governing documents. The Stamp Act, for example, transmuted into an assault on free expression, since it levied both commercial burdens and gatekeeping opportunities on all printed communications—hence, the eventual concern with protecting free speech, especially once England came down with prohibitions against citizen assembly, as it did in 1773.

Beginning in 1763, Parliment took it upon itself, for the first time, to impose taxes directly on the colonists, seeking to force the Americans to share the financial burden of the long and expensive French and Indian War. This did not go over well with the Americans, who had come to see "King Philip's War" as a struggle between two distant empires, fought out across the instrumentalized body of the innocent New World. One might argue, in fact, that it was Britain's dogged attempts to make Americans foot the bill for England's transference of Continental conflict to the North American sphere that provided the real impetus behind the American Revolution—and which supplied, in a roundabout fashion, much of the substance of today's constitutional conundrum.

Almost every clause in the Constitution can be linked to some specific act of oppression imposed on the colonists by the distant but determined British, beginning with appropriation of Indian and trade affairs by Crown appointees in the mid 1750s and continuing through to the infamous Stamp, Tea, and Intolerable Acts of 1765, 1772–1773, and 1773, respectively.

The much-debated Second Amendment to the Constitution—the right to bear arms—has a probable genesis in both the Intolerable Acts (1773), which gutted the Massachusetts Bay Colony charter and imposed unapologetic martial law in the region, and General Gage's actions in 1775, when he sent troops to confiscate arms and powder that had been secretly amassed by the colonials at Lexington and Concord. The Currency Act of 1764 denied Americans the right to print their own paper money, fueling, perhaps, the early nation's ferocity about fiscal independence.

The Quartering Act of 1765 forced them to host British officers in their own homes, at their own expense. Doubtless the colonial reaction to this outrage has persisted into today's national psyche, in the form of an enhanced concern for the protection of citizen rights, particularly privacy. The hated British Navigation Acts (1763) and the Revenue (Sugar) Act of 1764 gave the Crown a presumed right to dictate the terms of colonial business and trade—the colonial-era opposition to which, perhaps, laid the groundwork for today's peculiar American concept that "personal liberty" and unfettered free enterprise mean, essentially, the same thing.

All in all, 18th-century British arrogance has much to answer for in twentieth century struggles over the precise meaning and significance of constitutional intent. Even as they repealed the Stamp Act, in 1766, the British continued their bullying treatment of the mainland colonies by attaching to the repeal a bit of legislation (the Declaratory Act) seizing for the mother country the general right to make law in the Americas, even if they were thinking better of this particular law at hand.

Particularly odious to American colonials were the so-called "Intolerable Acts," imposed as punishment for the city of Boston's flagrant resistance to the uninvited tax on tea. Many residual notions, which came to life in response to these acts, continue to exist in the modern U.S. regulatory culture. They can be summed up, perhaps, in the general axiom that the the best government is that which governs the least—an appropriate response to the British ruling that closed the port of Boston, banned citizen assembly, permitted British troops to seize "vacant" buildings for their own use, sent erring government officials to Britian for trial (triggering the desire for an unbiased venue and a trial by peers), and, in general, imposing tyranny on a group of independent and resourceful souls.

What the colonial response to the Intolerable Acts failed to forsee, of course, was the eventual ascendency of industrial monopoly capitalism, a deficiency which is reflected in the design they drew up to keep the British nightmare from ever repeating itself on American soil.

Hickock (1991) and Kammen (1986) make the point, too, that many of the Constitutional framers opposed a Bill of Rights when it was first suggested, fearing that by listing the new nation's liberties, important freedoms might be inadvertently omitted, and thus lost. Several of the era's most prominant political thinkers felt that a Bill of Rights, within which the First Amendment is so snugly nestled, would be either redundant or hazardous.

George Washington, known more for his personal integrity than for his abstract political theorizing, was especially concerned with reconciling the interests of "diverse groups," and cautioned against confusing "individual liberty" with "social happiness." In this, he articulated the now underplayed concern with balancing citizen rights with citizen responsibilities— the idea of reciprocal obligation, or, as Kammen (1986) puts it, of "... (guaranteeing) freedom of action as long as it was not detrimental to others and was beneficial to the common weal" (p. ix).

Even James Madison, who somewhat reluctantly penned the first draft of the document, had to be prodded into doing so, primarily as a way to recoup on the loss of political capital incurred when he opposed Patrick Henry in Virginia state politics (Brubaker, in Hickok, 1991, pp. 82–93). Interestingly, Madison's first draft of the Bill had 17 clauses in it, 15 of which made it past the first draft, 12 of which made it into general committee debate, and 10 of which were adopted into the document we know today.

In his original preamble, Madison writes of a government's sacred duty to provide its citizens not just with liberty, but, indeed, with "safety" (Hickock, 1991, p. 4), a more assertive attitude toward government than is reflected in the surviving and intensely reactive Bill of Rights, which concerns itself, primarily, with stipulating what government *cannot do*, as opposed to being clear about what it *must* do. The distinction—an important one—is between a national blueprint concerned with negative rights and one which seeks to articulate positive responsibilities.

EXPANDING THE SCOPE OF CONSTITUTIONAL PROTECTION

The key idea, then, is that the remedy for a governing document which stubbornly persists in further marginalizing already-disenfranchised groups may lie not in developing new ways to argue against that document, but, rather, in shifting the original debate to new and different ground. The preceding examination of the cultural understructure of the U.S. Constitution suggests that reinterpreting the First Amendment may be an insufficient procedure for achieving social change, in that it bogs down in questions of procedure to the profound detriment of dealing with substance.

People get all caught up in how the First Amendment is worded, rather than in assessing whether it does what it is supposed to do—and whether or not it is the best way possible of achieving free speech objectives. It's almost as it we never go outside each individual constitutional clause to evaluate that clause's place and impact within the overall plan—as if each "right" stated in the Constitution is supposed to do its work in a moral vacuum, divorced from its peer liberties as well as from the society they are designed to encourage and protect.

Nor do we contemporary Americans, as a culture, factor into the constitutional equation the changed nature of media technology itself. The speed, the completeness, the idiosyncratic message coverage area, and the almost utter unaccountability of on-line communication demand a revisiting of the criteria by which public communication is guided. Yet this has yet to occur.

If it does nothing else, anything, the 1996 Telecommunications Act obscures what is a very pressing set of social concerns—the thought that there may be more to assessing the health and vigor of the increasingly on-line national press than a study of ownership patterns and rate structures. This ideological and historical decontextualization is specious, rhetorically, and dangerous, culturally, for it wedges apart rightfully connected aspects of the organic social whole. It's like trying to treat heart disease without acknowledging the presence of the hormonal system.

What may need to happen is that the Constitution, itself, may need to be extended—not altered, but enlargened, to include the more generalized principles necessary for adequate guidance in the complex welter of 21st-century civil liberties and human rights. Perhaps we need to think about adding some clauses that protect us from the threats provided by an urban, industrialized world, rather than from those anticipated by the citizens of an agricultural, newly postcolonial society. These new clauses would provide not just a more historically appropriate Constitution, but also a more equitable one, in that the increased universality of the guaranteed rights would make the overall document more responsive to modern economic conditions and social trends.

Is this, then, one way to ease the standoff in First Amendment interpretation—to circumvent the debate's interperative dimension altogther, by creating a legal instrument that can compete with the First Amendment at its own level of cultural resonance—by finding a way to fight fire with fire, so to speak, within the constitutional arena?

Perhaps. Given this deeply textured and richly complex historical legacy, what types of clauses, in the late 20th century, present themselves as candidates for a socially proactive version of the U.S. Constitution? Two come immediately to mind. The first of these is that which was articulated by an anxious Capitol Hill aide, who expressed the notion that: "People have a right to be safe." The second is an axiom that would supply some degree of general guidance to the intensifed confusion of intellectual property law on the electronic frontier: "People have a right to enjoy the benefits of that which they create."

FREE SPEECH VERSUS SAFE SOCIETY?

As part of a Congressional survey conducted in 1992, a number of Capitol Hill professionals were asked to identify what they considered to be the "single most important human right" (Harvey, 1992). A bevy of responses came back, generally clustered around the two themes "right to free speech" and "right to free enterprise." One respondent, however, added something that was not included on the original governing documents of the fledgling United States. This person reported that, in her opinion, the most important human right could be summed up in the statement: "People have a right to be safe." Interestingly, the right to be safe has been adopted as a regular clause in the 1996 South African bill of rights, given the same legal weight and cultural prominence as the freedoms of speech and religion.

In the United States, however, this situation is more complicated than it looks—and very, very important. While few scholars (or citizens) are likely to take issue with such a clear-cut and fundamental human right as personal safety, finding ways to achieve "safety" without infringing unduly on "freedom" is, essentially, the core dilemma of constitutional interpretation. As in the pornography issue, it can be argued that the operative dynamic comes down to the same set of normative questions: *Whose* safety versus *whose* freedom?

In terms of gender politics, the easy answer, of course, is that male freedom has been persistently framed as "more important" than female safety (although, according to the 104th Congress, not more important than children's safety). Independent of the patently reductionist nature of this explanation, though, is its failure to accommodate the historical complexities associated with changing cultural ideas surrounding the judicial process as a system for either challenging or defending the dominant status quo.

"People have a right to be safe." This seems like a basic and incontrovertible fact in a supposedly sane, civil, and advanced society. And yet, "safety," in the 20th-century United States, does not distribute evenly across lines of race and gender. Of course, precisely what is meant by the concept of "safety" is open to debate—security from measurable physical harm, protection from mental or emotional anguish, relief from social assault or psychic attack? Some of these aspects of "safety" are more tangible than others, and given the physicalized character of most U.S. law, those dimensions of safety which are more open to quantification are the most likely to enjoy protection under even as broad a social contract as this.

But even the most concretized of safety's attributes would challenge First Amendment absolutism, right from the start. For example, in the United States, there is at least one rape every 5 minutes. There is at least one murder every 22 minutes, at least one robbery every 49 seconds, and at least one auto theft every 19 seconds. All in all, according to the June 1991 report of the National Crime Survey, there is a violent crime happening, somewhere in the United States, once every 17 seconds, and the figures on violent crime seem to escalate yearly (U.S. Bureau of the Census, 1993, p. 189). These figures suggest that *something* is making the United States an especially hazardous place to live, and that, it can be argued, runs contrary to the "life, liberty, and pursuit of happiness" sentiments which triggered the entire Constitutional event in the first place. The British, it seems, are no longer the threat; we have become dangerous to ourselves.

Of course, the job of providing "safety" to all Americans will, eventually, put the rights of some Americans on a collision course with the rights of other Americans. Impact is inevitable. And there is always the danger—the very real danger—that deliberately restrictive forces within the political sphere might find a way to pervert the intention of the "safety" phrase in attempts to further marginalize, if not outright vilify, those groups and individuals within the larger culture that they may find personally objectionable. But it may be worth the risk, since demonization seems to have been a strategic presence in American politics long before this particular suggestion about Constitutional amending was ever raised.

The question presents itself, then, of whether we should consider seriously the addition of an amendment to the U.S. Constitution, similar to the one included in the South African parallel document. The prime utility of a safety amendment suggestion, perhaps, lies in the fact that crucial legal distinctions are achieved only when colliding legal bodies carry similar degrees of constitutional clout. Pitting one constitutional amendment against another—the First Amendment right of an on-line pornographic magazine against the social safety rights of an at-risk community, for example—gets around the David-and-Goliath aspect of trying to stop a steamrolling First Amendment with anything less than its legal equal. Everything pales before constitutionally guaranteed rights; all remedies against socially inequitable interpretations of the constitutional spirit seem puny when confronted with the culturally sacred wording of the original document. In essence, the idea, here, is to fight fire with fire, rather than to beat at it feebly with a wet blanket.

A secondary charm of the safety amendment is that it does not confine itself to governmental intervention. It would apply with equal force to private organizations and corporate institutions, thus in a very real way sidestepping the charge that it violates free speech restrictions altogether, since the relevant target of the First Amendment is limited, by definition, to the government.

The point of all this legal hair-splitting, it should be noted, is not to gut the spirit or subvert the intention of the First Amendment, but to complement it with other equally important, albeit superficially antagonistic, human rights protections. There is some precedent for this. For example, a history of creative tension between the First and Sixth Amendments has yielded a relatively viable approach to the free press/fair trial dilemma, providing both the latitude and the elasticity needed to make intelligent case-by-case decisions without lapsing into procedural idiosyncracy. Adding "social safety" as a proactive liberty in its own right, rather than treating it as a collateral benefit of other constitutional protections, would supply a similarly general promise that would be adaptable to specific case-by-case applications. Were such an amendment in place, for example, Laurie Powell and Rebecca Shaeffer might have very different stories to tell today.

CONCLUSIONS

The constitutional framers probably never anticipated a world as diverse and as chaotic as the one occupied by 20th-century Americans. Certainly, neither could they have imagined the fantastic and profoundly revolutionary technologies which would, eventually, define the social parameters of that world. But the governmental tools they left behind are, nonetheless, still viable models for ordering human affairs in a relatively equitable fashion.

The tool kit is not so much inadequate as it is incomplete. And that can be remedied. But the place to start is with the realization that one set of wrenches, for example, is not intrinisically more sacred than another, and that sometimes the right tools for any given job must be invented by the people who actually need to use them. There is no shame in that sort of ingenuity, only an invitation to bold and creative thinking...and a great opportunity for a rare sort of architectural satisfaction. We Americans, like everybody else, must live in the world as we find it—on-line and off line, both. But nothing except the inherited limits of our own cultural vision keep us from improving the social blueprints which guide it, as we go along. On line, as well as off.

REFERENCES

Bagdikian, B. H. (1987). *The media monopoly* (2nd ed.). Boston: Beacon Press.

Bennett, W. L. (1983). *News: The politics of illusion.* New York: Longman.

Borden, D. L. (1993, October 6). *Women and the First Amendment.* Remarks from the Radford University Forum on Women and the First Amendment, Radford, VA.

Bowles, S. (1996, January 26). Police search of AOL files divides the on-line world. *The Washington Post,* p. A1.

Bowles, S., & Shields, T. (1996, February 15). Internet message prompts call for a student code. *The Washington Post,* pp. B1, B6.

Byerly, C. (1993, October 6). Opening remarks, Radford Forum on Women and the First Amendment, Radford University, Radford, VA.

Cate, F. H. (1992). Introduction: The First Amendment and problems of constitutional line-drawing. *Visions of the First Amendment for a new millennium.* Washington, D.C.: Annenberg Washington Program.

Cavazos, E. A., & Morin, G.. (1994). Cyberspace and the law: Your rights and duties in the online world. Cambridge, MA: MIT Press.

Dates, J. L. (1991). "Multiculturalism and the broadcast curriculum," *Feedback,* pp. 4–13.

Dickson, T. D. (1993, Winter). "Sensitizing students to racism in news." *Journalism Educator, 47*(4), 28–33.

Downs, M. (1996, February). *Internet Underground,* pp. 35–36.

Fishman, M. (1980). *Manufacturing the news.* Austin: University of Texas Press.

Gitlin, T. (1980.) *The whole world is watching.* Berkeley, CA: University of California Press.

Harvey, K. (1992). *Blueprints for a presidency.* Unpublished manuscript.

Hickock, E. W. (Ed.). (1991). *The Bill of Rights: Original meaning and current understanding.* Charlottesville and London: University of Virginia Press.

Johnson, O. (Ed.). (1992, June). National Crime Survey Report. *Information Please Almanac.* (45th ed.) Boston: Houghton Mifflin Company.

Kammen, M. (Ed.). (1986). *The origins of the American Constitution.* New York: Penguin Books.

Kim, J. (1995, November 13). Business bet on the future: despite array of products, there's no winner—yet. *USA Today,* pp. 1E, 2E.

Kornbluth, J. (1995, December 24). Who needs America Online? *New York Times Magazine,* pp. 16–19, 36, 40.

Lang, G. E., & Lang, K. (1983). *The battle for public opinion: The president, the press, and the polls during Watergate.* New York: Columbia University Press.

Mark, M. E. (1995, January). Terror on-line. *Vogue,* pp. 160–166, 195.

Newcomb, H., & Hirsch, P. (1987). "Television as a cultural forum." In Horace Newcomb (Ed.), *Television: The critical view,* (4th ed.), pp. 455–504. New York: Oxford University Press.

Powers, W. F. (1994, November 8). Virtual politics: Campaigning in cyberspace. *The Washington Post,* pp. E1–E2, E6.

Progress and Freedom Foundation. (1994, August). Cyberspace and the American Dream: A Magna Carta for the Knowledge Age. *Future Insight, 1*(2). Available from 1250 H St. N.W., Washington, D.C., 20005.

Telecommunications Act of 1996. 104th Cong., 2d Session. (1996, January 31). House of Representatives Report 104–458: Washington, DC.

Times-Mirror Center for the People and the Press. (1994, May 24). *Technology in the American Household.* Available from 1875 Eye St., N.W., Suite 1110, Washington, D.C. 20006.

Tuchman, G. (1978). *Making news: A study in the construction of reality.* New York: The Free Press.

U. S. Bureau of the Census. (1993). Section five, pp. 189–214. *Statistical abstract of the United States: 1993.* (113th ed.). U.S. Government: Washington, DC.

Cyberlibel:
Time to Flame the Times Standard

Diane L. Borden
George Mason University

INTRODUCTION

The message on the computer bulletin board based its charge on "rumors." It alleged that a university professor had engaged in pedophilia. It also charged that he had no genuine academic ability, that he was a racist, that he was a tool of big business, and that he drank to excess (*Rindos v. Hardwick*, 1994).

The message made its way through the electronic grapevine, in this case the "sci.anthropology" bulletin board, in June 1993. The professor, archaeologist David Rindos, recently had been denied tenure; he brought a libel action against the sender of the message, professor Gilbert Hardwick, claiming that the bulletin board material, posted on the Internet, was defamatory. The court rejected three of the claims but agreed that the imputations of sexual misconduct and lack of professional competence were seriously defamatory. Hardwick failed to appear in court to defend his action, and the Australian court awarded Rindos $40,000 (about $28,000 US) in damages, satisfied that he had suffered reputational harm. The case was widely believed to be the first in the world in which a court rendered a judgment for a plaintiff for cyberlibel.

In the United States, only a handful of on-line defamation cases have been filed, and, for the most part, the plaintiffs have been nonmedia corporations. Of the most notorious cases, two were settled before trial (*Medphone v. Denigris*, 1992, and *Suarez v. Meeks*, 1994); one was decided in a state court (*Stratton Oakmont v. Prodigy*, 1995); and one was decided in a federal district court (*Cubby v. CompuServe*, 1991). Unlike traditional libel cases, in which the *plaintiff's* status in the community is central, the issue in the on-line cases seemed to have turned on the *defendant's* status as either a "publisher" or a "distributor" of information. In both

traditional and on-line defamation actions, issues about the nature of the defamatory statements, their truth or falsity, or the harm to the plaintiff's reputation are
almost always shoved to the background.

At one time, protecting reputation was at the heart of defamation law. Keeton
et al. (1984), Pollock (1894), and others in the legal sphere have extolled the individual's interest in maintaining a good reputation, likening the importance of its
protection to that afforded bodily safety and freedom. Laws were passed to punish
those who caused reputational harm.

When the U.S. Supreme Court constitutionalized defamation law in its *New
York Times v. Sullivan* (1964) ruling, however, it shifted the weight of the law's
scrutiny from concern over reputational harm and truth or falsity of the defamatory statement to the motivation of the media defendant: Was the defamatory
statement published with actual malice? In so doing, the Court also shifted the
burden of proof from the defendant to the plaintiff and set the stage for a decades-
long definitional battle over the concept of "publickness." In *The New York Times*,
the Court created a fault standard that turns on what constitutes "public" speech
and what constitutes a "public" plaintiff. The *Times* decision and its progeny also
have left open the question of whether the First Amendment offers a different
kind of protection for media defendants than it does for nonmedia defendants.

This chapter argues that the *Times* standard is obsolete and that efforts to reform defamation law, refocusing on remedying reputational harm, need to be
dusted off and revisited in this new age of cyberspace, cybercommunications, and
cyberlibel.

As of this writing, no court has rendered a judgment in a case brought against
a traditional mass medium (newspaper, magazine, broadcast station) for on-line
defamation, probably because so few media have on-line operations. Only about
300 commercial newspapers worldwide, compared to the 10,000 such products
available, have on-line operations; in addition, only about 5,000 magazines,
newsletters, newswires, and television and radio transcript services have on-line
operations, compared to the 100,000 such products available in print. But the burgeoning use of computer-based communications technologies most certainly will
change the contours of the debate over libel fault standards, particularly the definitions of publickness and, in so doing, will revolutionize the law of defamation.

DEVELOPMENT OF "PUBLICKNESS" STANDARDS

To understand how developing communications trends will shape the future of
defamation law, it is necessary to review how the law has settled into its current
state. The path has been tortuous, and today's law is a bewildering study in confusing doctrine and jurisprudence fragmented by conflicting beliefs and opinions.
Keeton et al. (1984) hit the nail on the head when they wrote:

[T]here is a great deal of the law of defamation which makes no sense. It contains anomalies and absurdities for which no legal writer ever has had a kind word, and it is a curious compound of strict liability imposed upon innocent defendants, as rigid and as extreme as anything found in the law, with a blind and almost perverse refusal to compensate the plaintiff for real and very serious harm. The explanation is in part one of historical accident and survival, in part one of the conflict of opposing ideas of policy in which our traditional notions of freedom of expression have collided violently with sympathy for the victim traduced and indignation at the maligning tongue. (p. 772)

Particularly bewildering are the amorphous distinctions courts have made between public figures and private individuals, between statements of fact and opinion, and between issues of public concern and those of purely private concern (see discussion of the *Times* opinion below).

A. Common-law defamation

In the United States, all laws concerning defamation are enacted by the states, and people who believe they have been falsely defamed may bring a cause of action, usually in a civil court, seeking monetary damages. Criminal libel laws still do exist in some states, but they are rarely invoked (Tedford, 1993, p. 85).

Communication is defamatory if it "tends to harm the reputation of another as to lower him [*sic*] in the estimation of the community or to deter third persons from associating or dealing with him [*sic*]" (Keeton et al., 1984, p. 774). Defamation may take two forms: Slander pertains to spoken defamation, libel to written defamation. The transmission of defamatory messages in cyberspace more closely resembles libel than slander. Both slander per se and libel per se apply to four types of statements, which are "defamatory on their face—that is, they are considered to be intrinsically, automatically destructive of reputation" (Keeton et al., 1984, p. 774). These four types of statements include those that impute a crime; those that impute a loathsome disease; those that are injurious to a plaintiff in his business, trade, profession, office or calling; and those that impute immorality (Keeton et al., 1984, p. 788).

Before 1964, the common law's interest in protecting reputation was one of strict liability, requiring only that the plaintiff establish defamatory communication (i.e., that the defamation was intentionally communicated by the defendant to a third person). The defamatory statement was viewed as false unless the defendant proved truth, or called upon a complex structure of privileges, such as reporting on public meetings or commenting on matters of public concern. The issue of fault was irrelevant.

B. Constitutionalization of libel law

Since *New York Times v. Sullivan* (1964), the U.S. Supreme Court has determined that a consideration of fault **is** relevant, and that in defamation actions against the mass media, plaintiffs now must demonstrate motivation on the part of the defendant. The *Times* case involved a full-page advertisement in the March 29, 1960, issue of *The New York Times.* Headlined "Heed Their Rising Voices," and signed by the Committee to Defend Martin Luther King and the Struggle for Freedom in the south, the ad protested a "wave of terror" against Black people involved in nonviolent demonstrations in the south.

The plaintiff was an elected commissioner in charge of the police department in Montgomery, Alabama, who sued the newspaper for libel. Sullivan was not named in the ad, but Alabama law provided that criticism of the department he supervised reflected on his reputation. Alabama's libel law also provided for strict liability (see explanation above) and for truth as an affirmative defense. Although the ad contained several factual errors (e.g., Dr. King had been arrested only four times, not seven, and the student protesters sang "The Star-Spangled Banner," not "My Country 'Tis of Thee"), and state law required a higher standard than carelessness in the award of punitive damages, the Alabama courts awarded the plaintiff $500,000 in damages.

The U.S. Supreme Court reversed the judgment and offered a constitutional privilege to media libel defendants to publish and disseminate information of public concern. (The Court has vacillated on the issue of constitutional protection for nonmedia libel defendants.) In *New York Times*, the Court held that the First Amendment guaranteed "uninhibited, robust, and wide open" debate on public issues and that, in the interest of such debate, public officials, such as Sullivan, could not recover for defamatory falsehoods unless they proved the statements were made with actual malice, defined as "knowledge that [the statement] was false or with reckless disregard of whether it was false or not" (*Times v. Sullivan*, 1964, pp. 279–280). The standard is based on the notion that public officials have greater access to the media than ordinary citizens and can, therefore, command a right to reply to the defamatory statements and to offer a public rebuttal.

The concept of wide open and robust public debate is the central theoretical framework of the decision and is based on the work of political scientist and philosopher Alexander Meiklejohn, who argued that in a democracy, a properly informed electorate must hear all relevant ideas and opinions in the marketplace. Speech concerning the governmental process, therefore, deserves absolute protection from governmental interference. Nonpolitical speech, less important to the affairs of state, may be regulated or abridged by due process of law (Meiklejohn, 1979).

In later cases, the Supreme Court extended the actual-malice standard so that proof of differing degrees of fault is now required in libel actions. The determination of the degree of fault turns on two factors: the plaintiff's status in the community and the nature of the speech itself (i.e., whether it is defined as public speech or private speech. Each of these two fault standards is examined below).

C. Difficulties in definitional boundaries

The judicial system, plaintiffs, and defendants have wrestled with the often conflicting and ambiguous standards of "publickness" for more than three decades, and the courts have unraveled as many definitional strands as they have woven.

STATUS OF PLAINTIFF. In terms of the plaintiff's status in the community, the actual-malice standard applies not only to *elected public officials*, such as Commissioner Sullivan, but also to *nonelected public officials*. Two years after the *Times* decision, the Court defined public officials, generally, as those who have substantial responsibility within the government hierarchy (*Rosenblatt v. Baer*, 1966). The plaintiff in the case was the supervisor of a public recreation facility in New Hampshire. He brought a libel action against a weekly newspaper, which reported that the facility was financially more sound a year after his discharge than it had been during his tenure. The Court held that plaintiff Baer was a public official and was subject to proof of actual malice, reasoning that his standing in the community would give him *access to the media* and he could, therefore, rebut any defamatory accusations hurled at him. "It is clear. . .," the Court held, "that the 'public official' designation applies at the very least to those among the hierarchy of government employees who have, or appear to the public to have, substantial responsibility for or control over the conduct of government affairs" (*Rosenblatt v. Baer*, 1966, p. 85).

In addition to nonelected public officials, the Supreme Court extended the actual-malice standard to *public figures* in two companion cases 1 year later. In *Curtis Publishing Co. v. Butts* (1967) and in *Associated Press v. Walker* (1967), the Court found that "some people, even though they are not part of the government, are nonetheless sufficiently influential to affect matters of important public concern" (*Associated Press v. Walker*, 1967, p. 164). Again, the definitional presumption is that public figures have access to the media and are able to command a right to reply to the defamatory statements. In *Butts*, the plaintiff was an athletic director at a state university and was accused of disclosing a game plan to an opposing coach. The Court deemed him a public figure. Similarly, in *Walker*, the plaintiff was a retired Army general reported to have led students in an attack on federal marshals enforcing a desegregation order. The Court also deemed him a public figure.

But in three other cases only a decade later—*Time, Inc. v. Firestone* (1976), *Hutchinson v. Proxmire* (1979), and *Wolston v. Reader's Digest Ass'n* (1979)—the Court determined that the plaintiffs were **not** public figures and therefore did not have to demonstrate actual malice on the part of the defendant. In *Firestone*, the plaintiff was a socialite involved in an 18-month divorce trial that attracted national media attention. In *Hutchinson*, the plaintiff was a scientist and professor at a state university, who had successfully solicited $500,000 in federal research grants. In *Wolston*, the plaintiff was a man convicted on contempt-of-court charges for refusing to testify before a grand jury investigating Soviet espionage. The "publickness" distinctions among all these cases remains elusive.

Yet, at about the same time, the Court pushed the definitional boundaries of who constitutes a public figure even further and, in *Gertz v. Robert Welch, Inc.* (1974) required that *private-figure* libel plaintiffs demonstrate a minimum level of fault, such as negligence, in order to recover actual damages, and that they demonstrate actual malice—that the defendant acted with knowledge of falsity or in reckless disregard of the truth—in order to recover punitive damages.

Gertz was an attorney retained to represent the family of a youth killed by a Chicago police officer in 1968. He was drawn into a public campaign by the John Birch Society to publicize an alleged nationwide conspiracy to discredit local police. The society's magazine, *American Opinion,* published a story critical of Gertz, portraying him as a "major architect" in the conspiracy and falsely asserting that he was an official of the Marxist League, a Leninist and a "Communist-fronter." In its opinion, the Court declared Elmer Gertz a private figure but required, for the first time, that the plaintiff show negligence on the part of the defendant in order to recover actual damages.

In *Gertz*, the Court attempted to broadly define public figures as (a) those who have assumed roles of great prominence or power in society; or (b) those who have thrust themselves in the vortex of a public controversy. Private individuals are those who have not assumed an influential role in the public eye. But, in requiring even a minimum fault standard of private individuals, Justice White noted in his dissent, the Court radically change the common law of libel, shifting the burden of proof to the private-figure victim, "even though he has done nothing to invite calumny, is wholly innocent of fault, and is helpless to avoid his injury" (*Gertz v. Welch*, 1974, p. 390).

Justice White added:

> To me, it is quite incredible to suggest that threats of libel suits from private citizens are causing the press to refrain from publishing the truth. I know of no hard facts to support that proposition, and the Court furnishes none. The communications industry has increasingly become concentrated in a few powerful hands operating very lucrative businesses reaching across the Nation into almost every home. Neither the industry as a whole nor its individual components are easily intimidated, and we are fortunate they are not. Requiring them to pay for the occasional damage they do to private reputation will play no substantial part in their future performance or their existence. (*Gertz v. Welch*, 1974, pp. 390–91)

STATUS OF SPEECH. The second factor judged necessary in finding a cause for action in libel is whether the speech at issue is of public or private concern. Embracing the ideas expounded by Meiklejohn and others, the *Times* decision made clear that debate over issues of public concern (i.e., political speech), should receive the strongest constitutional protection. But later cases expanded the "publickness" of speech to include broader societal issues: "[T]he guarantees for speech and press are not the preserve of political expression or comment upon public affairs, essential as those are to healthy government," the Court argued (*Time v. Hill*, 1967).

In the mid-1980s, the Court set about to structure a hierarchy of speech values (*Dun & Bradstreet, Inc. v. Greenmoss Builders, Inc.*, 1985) in order to more clearly determine the weight of constitutional protection required to balance society's interest in wide-open debate against the three tiers of individual interest in reputation (public official, public figure, private individual) set out in *Gertz* (1974). The speech pyramid, the Court reasoned, requires the strongest constitutional protection for speech of public concern at the apex and demands the least constitutional protection for speech of purely private concern at the base of the triangle. Some have argued that this distinction is gender-specific (Borden, 1997) and is constructed around cultural distinctions between the private sphere and the public sphere. Nevertheless, the Court held that a private plaintiff could recover actual and punitive damages for a defamatory falsehood, without a showing of actual malice, if the speech were not a matter of public concern (*Dun & Bradstreet v. Greenmoss*, 1985). A year later, however, the Court attempted to distinguish public speech from private speech even more and held that, contrary to common-law presumptions, the burden of proving the falsity of speech of public concern is upon the plaintiff, even when the plaintiff is a private individual (*Philadelphia Newspapers, Inc. v. Hepps*, 1986).

As a result of the *Times* ruling and its progeny, the scales of justice have been tipped squarely in favor of media defendants. The courts have found that the public's interest in a free press outweighs an individual's interest in protecting reputation. As a result, studies indicate, plaintiffs ultimately prevail in only 5% of all libel actions brought against the mass media (Franklin, 1980, p. 479).

It appears, then, that the only remaining category of speech that addresses reputational harm is a cause of action for the publication of a **defamatory falsehood** (meaning false facts that cause reputational harm, only; opinion holds absolute constitutional protection) aimed at **private individuals** concerning purely **private concerns**, which the courts have not yet defined.

CYBERSPEECH AND ON-LINE DEFAMATION LAW

On-line communication is an amalgam, a hybrid of various technological forms that preceded it, which has swept into its fold millions of users. In 1994, an estimated 5.2 million people logged on to on-line services, up from 1.7 million only

4 years earlier. In 1995, an impressive 24 million North Americans regularly logged on to the Internet (Levy, 1995), that amorphous worldwide network of computers born 30 years ago under the auspices of the U.S. Defense Department.

In some respects, the burgeoning on-line world resembles its precursors: It is similar to newspapers in that "subscribers" pay to read its messages. This type of on-line service—LEXIS or NEXIS are examples—distributes information but does not allow interaction with the receiver. In other ways, on-line communication "sounds" more like a telephone conversation or a radio talk show since "listeners" (users) may respond to a message immediately and spontaneously. This type of service—Prodigy, CompuServe, and America Online are examples—both provides information and allows users to interact with groups of other users through bulletin boards. Finally, on-line communication can be much like a personal letter to a family member or a friend, or a memo to a business colleague. This type of service—called electronic or e-mail—sends private messages to other individuals through a central computer. On-line communication also can be all of these things at once.

But one distinguishing characteristic of travelers in such a cyberspatial world, particularly as it relates to defamation law, is the facility and ease with which they may make themselves anonymous or may transform themselves into someone they are not—at least someone they are not in the physical world.

A. Current on-line case law

A 1995 on-line defamation case involved such an anonymous message and, as *Business Week* reported, "brought to a head a long-anticipated clash between traditional law and freewheeling computer communications. . . .The basic problem: Existing laws are outdated for today's direct, real-time communications" (Yang, 1995).

The case involved an anonymous user of the interactive on-line service, Prodigy, who posted three messages on "Money Talk," an electronic bulletin board, in October 1994. The messages accused investment banking firm Stratton Oakmont Inc. and its president of criminal fraud and dishonesty. Stratton Oakmont's response was to bring a $200 million libel suit against Prodigy and the anonymous flamer, alleging that the message had caused the firm humiliation and public disgrace throughout its on-line network (*Stratton Oakmont v. Prodigy*, 1995).

The issue in this case turned not on reputational harm, nor on the truth or falsity of the defamatory statements. The issue was whether the on-line service Prodigy was a "publisher" of information or simply a "distributor" of information. The liability standards are quite different for each.

Four years earlier, the courts had defined another on-line service provider, which had been sued for libel, as a "distributor" of information. In *Cubby v. CompuServe* (1991), the court held that CompuServe had little or no editorial control over the content of messages on its bulletin boards and that, as a result, it should be deemed a "distributor" (or secondary publisher) of news—more like a library or a newsstand than a newspaper. It, therefore, should be absolved from liability for defamatory speech, unless it distributed such material knowing it was false or with reckless disregard of its truth.

In contrast, the court found that Prodigy clearly employed strict content controls by utilizing software which screened its system for taste and etiquette (including obscene language and racial slurs) and by relying on system operators (editors) to screen posted messages. The court, therefore, ruled that Prodigy should be elevated to the status of a primary publisher. But, because the case eventually settled out of court with Prodigy's apology to Stratton Oakmont, the court had no opportunity to define a standard of liability.

Two other cyberspace defamation cases in the United States have been settled out of court: *Medphone v. Denigris* (1992) and *Suarez v. Meeks* (1994). In 1992, Peter Denigris logged on to Prodigy's "Money Talk" bulletin board and posted a message urging investors to be wary of the Medphone Corp. because the company appeared to be engaged in fraudulent practices. Medphone responded by bringing a libel action in the federal district court in New Jersey, arguing that Denigris' comments reduced the price of the company's stock, damaged its reputation, and caused the company irreparable harm. Before a court could decide the matter, however, the parties agreed to settle the matter out of court. Medphone, which had sought $40 million in libel damages, agreed to drop the action for $1 and a promise from Denigris that he would not make any future false statements about Medphone via Prodigy or any other means of communication (*Medphone v. Denigris*, 1992).

In the second case, which some observers considered the first libel suit over an Internet transmission, Brock R. Meeks, editor of an electronic newsletter called *Cyberwire Dispatch*, posted two messages in 1994, accusing Suarez Industries of running an electronic postal service scam. Through on-line research and follow-up confirmation through authorities, Meeks discovered that the state of Washington had sued Suarez for similar violations of various consumer protection laws and had enjoined Suarez from continuing its mass mailing solicitations. Meeks posted these findings in his Internet newsletter.

Unlike CompuServe, Prodigy, and other on-line service providers, the Internet has no central operator to hold liable for libelous remarks. Suarez, therfore, sued Meeks for defamation directly, alleging that two portions of the articles were false and defamatory. The two statements were: (a) "Let's flip this latest Internet scam on its back and gut that soft white underbelly," and (b) "He's (Benjamin D. Suarez) infamous for his questionable direct marketing scams. And he has a mean streak. His record speaks for itself."

In a motion for summary judgment, Meeks argued that Suarez should be subject to the *Times* fault standard: proof of actual malice. Meeks argued that Suarez was a public figure not because of his position in society but because, through his company's mass mailing solicitation on the Internet, he voluntarily exposed himself to scrutiny by the Internet public. The motion declared:

> Through their solicitations alone plaintiffs assumed the risk of comment and criticism about their business activities, but they assumed an enhanced risk of criticism by selecting the Internet as the medium for their communications. As any Internet user knows, the Internet is a free-for-all, no-holds-barred communications medium. "Flaming," or caustically criticizing those with whom one disagrees, is a common reaction to Internet postings. (*Suarez v. Meeks*, 1994, p. 28)

In addition, Meeks argued, like any other Internet user, Suarez had ample opportunity to respond to the criticism by posting a reply rather than suing for libel.

Before the court could rule on the motion for summary judgment, however, the parties settled out of court. Meeks claimed legal fees to defend himself had grown too burdensome, and he agreed not to publish any article about Suarez for 18 months without allowing Suarez the opportunity to review the article first.

B. "Publickness" standards irrelevant in a cyberworld

If the cyberspace defamation cases all involved defendants who were the primary publishers of the material, the first step the courts would take under current law would be to apply the two standards of publickness examined above: whether the plaintiff is a public official, public figure, or private figure; and whether the speech at issue is of public concern or private concern. The answers to these questions would determine the plaintiff's standard of fault: actual malice or negligence.

In an on-line universe, however, even the fuzzy distinctions the courts have made between public and private figures vanish, and the distinctions between public and private speech become irrelevant. If, as Justice Powell noted, the "first remedy of any victim of defamation is self-help—using available opportunities to contradict the lie or correct the error and thereby to minimize its adverse impact on reputation" (*Gertz v. Welch*, 1974, p. 390), then users of on-line communications have a more realistic opportunity to counteract on-line defamation than private individuals currently have to rebut defamation published in the traditional mass media, to which they have little or no access.

For all intents and purposes, users of cyberspatial communication media could all be defined as "public" people since, by the very nature of their interactive medium, they have a forum for their own speech as well as the means to reply to a defamatory falsehood published by another. Victims of defamation may respond to the besmirching of their reputations by getting online, tapping the "reply" button on the keyboard, and sending their message to the same list of users to which the defamatory statement was distributed.

In addition, cyberspace users have the ability to create their own on-line personae, which may be as distinct from their real selves as the Internet is from the
physical world, and, to a certain extent, they have the ability to control how they
are perceived by others:

> The degree to which Usenet users resemble their personae seems to vary.
> The representation of a user within Usenet is the attempted transfer of the
> user's individuality into a Usenet persona. The user has some control
> over the representation and the extent to which the persona resembles
> himself or herself. (MacKinnon, 1995, p. 118)

This ability to create artificial or fictional personae online may prove to be of
grave concern in the area of libel law. Given the case law to date, real victims of
on-line libel may have difficulty determining the source, or publisher, of the libel.
If the defamatory statements occurred on a bulletin board service without editorial
controls (see *Cubby v. Compuserve* discussion above), for example, and the source
of the libel were a fictional persona, then the libel victim would have little recourse
beyond a personal reply. For some, however, a right of reply is only a small part
of the necessary recovery of one's good name.

REDISCOVER NOTION OF REPUTATIONAL HARM IN CYBERSPACE

In the 30 years since the Supreme Court's *New York Times* ruling, the dissatisfac-
tion with defamation law has grown, leading to a series of reform proposals based
on the premise that the law is in such a state of disarray that it fails on at least three
fronts: It does little to protect reputation or redress reputational harm; it does little
to help determine the truth; and, even as it offers the media almost absolute privilege to defame, when litigation does arise, the financial costs to both plaintiffs and
defendants are staggering. The parallel growth during the last three decades of on-
line communications and the potential they offer not only for "robust" civic debate
but also for harm to individual reputation, provides a fortuitous opportunity for a
reconsideration of the central purpose of defamation law.

A. Add weight to reputational harm

The concept underlying laws to punish defamatory expression is that each person
has a right to protect his or her reputation from harm. In the United States, laws
concerning defamation are enacted by the states, and people who believe they have
been falsely defamed may bring a cause of action, usually in a civil court, seeking
redress, including monetary damages. While each state's defamation law varies
slightly in language and in date of enactment, most suggest that "speech can be
called defamatory if it tends to lower a person's *reputation* before others, cause

that person to be shunned, or expose that person to hatred, contempt, or ridicule" (Tedford, 1993, p. 84).

Despite the law's forthright claim that reputational harm requires remedy, however, modern courts rarely engage in consideration of the plaintiff's reputation. In addition, little has been written in either the social histories or in the legal literature about reputation, per se. Perhaps that is because the concept is such an elusive one.

Reputation, one observer suggests, is as mercurial as speech itself: "The vulnerability of a good name stems from the fact that it is held and conferred by people other than the person who is said to possess it, and that it has no tangible substance, it consists entirely of words" (Wilson, 1974). Yet, he observes, the concept of reputation presupposes how people relate to one another in their social roles: "Each individual's good name is defined by others according to shared and commonly sanctioned criteria of interest" (p. 100).

Another argues that defamation law affords protection to the "individual's projection of self in a society" (Skolnick, 1986, p. 677). Yet another suggests that, although language insists that people "own" or "have" a reputation, "reputation is not a property or a possession of invidivuals—it is a relation between persons" (Bellah, 1986, p. 743). Probert (1962) argues that the law's traditional assumptions hold that a person's reputation is comprised of what other people think of him and what image they hold of him (p. 1176). Norton (1987) suggests that a person's standing in a community depends on his or her reputation, and "reputations are sustained or lost primarily through gossip" (p. 5).

While widely varying in their approaches, the commonality among these concepts is the presumption that reputation has something to do with the ways in which people relate to each other in their social roles. In other words, an individual's good name is defined by others, by society, by the culture at large.

The Langs note that because people value their reputations so highly and because reputations are so difficult to earn, that, once earned, they are entitled to legal protection:

> As regards one's own reputation, most people consider it among their most prized possessions, even though it cannot, like material goods, be inherited or directly passed on to someone else. Reputations have to be earned and validated, but once established, they are entitled to legal protection, much like other property, so that a person whose reputation is damaged through malice or negligence can seek redress through the courts. (Lang & Lang, 1990, p. 6)

The Langs posit, however, that reputation may not be entirely left to the whims of society, and that, for one, the artistic reputation, just as the artistic product, can be created by the individual:

> The posthumous reputations of visual artists are more specifically linked to surviving examples of their creations; neither documents testifying to their achievements nor authenticated copies suffice. Nothing can substitute for the originals, for objects wrought by the artists' own hand. Consequently, so runs our argument, what artists do in their lifetime to facilitate the preservation and future identification of their oeuvre has a significant effect on whether, and how well, their names will be known to posterity. (Lang & Lang, 1990, p. 317)

But, whether posthumously or in the flesh, the law has failed to adequately define the concept of reputation, despite its supposition that harm can come to it through the power of speech or of the written word.

Post argues that by categorizing the nature of the harms suffered, one may discover the kinds of reputation the law considers deserving of protection. He argues that the common law of defamation has attempted to protect a *man's* reputation (see comment below about the genderization of reputation) as property, as honor and as dignity, and that each "corresponds to an implicit and discrete image of the good and well-ordered society" (Post, 1986, p. 693). Reputation as property, he suggests, assumes that one's reputation can be earned through diligence and hard work and that the law will protect the market value of that reputational "product." Conversely, reputation as honor assumes that one's reputation is not earned but adheres by right of the status granted the person's social role, such as a king or a gentleman. As such, reputation as honor presupposes that individuals are unequal because they perform different social roles. Reputation as dignity is a more complicated concept, based on the twin beliefs that an individual has an interest in being included in the forms of social respect at the same time that the society has an interest in defining and maintaining the rules of cultural identity (Post, 1986, pp. 694–695, 700, 710–711).

As intriguing as Post's analysis is, absent from his work and from the literature in general is an analysis of harm that may come to a woman's reputation (Norton's 1987 study of 17th century defamation in Maryland is a rare exception). Another recent study attempts to shed light on reputational harm for women and suggests that such harm is historically and culturally related to the twin concepts of separate spheres and the virtuous woman. The study suggests that the law created and continues to reinforce separate, gendered definitions of reputational harm: for a man, it is defamation of his honor; for a woman, it is defamation of her sexual virtue (Borden, 1997).

Some legal observers have suggested that the courts return to a consideration of the core idea embodied in defamation law: People have a right to protect their reputations from defamatory falsehood. Recognizing that reputation is socially determined, Anderson argues that plaintiffs should be required to prove reputational harm and suggests four distinct ways to do so:

- By claiming that the defamation may interfere with the plaintiff's *existing* relations with other people;
- By claiming that the defamation may interfere with the plaintiff's *future* relations with other people;
- By claiming that the defamation may destroy the plaintiff's favorable public image;
- By claiming that the defamation may create a negative public image where no public image existed previously (Anderson, 1984, pp. 765–766).

In cyberspace, reputation is also significant. In fact, some argue that reputation is the most important benefit of power in cybersociety:

> . . .[O]ne's powers, such as strength and eloquence, are expressed by participating in the cycle of statements and responses. Only in this way can one's powers be perceived, substantiated, measured, and ranked by others. The resulting comparisons made among personae establish the public estimation of one's worth. . . .[R]eputation is the collective memory of the comparisons of past cycles of statement and response. (MacKinnon, 1995, pp. 126–127)

Unlike the traditional mass media, however, on-line communications are relatively unmediated, and the resulting freedom to speak whatever is on one's mind often leads to flame wars, the hurling of argumentative and inflammatory statements, often designed for hastened reputational gain. Online, this source of conflict is called "baiting" or "posting flame bait." Often, the practice involves the deliberate posting of a message that is a clear violation of cyberspatial conventions; or it may involve posting a message that tries to "sucker" as many people as possible. Sometimes, as a way to draw unwanted attention to another and to cause that person reputational harm, the poster may send the message from someone else's account. The post being replied to below probably was an attempt to bait gullible readers (Merchant, 1994):

From: [name deleted]

Newsgroups: rec.pets.cats, alt.syntax.tactical, alt.flame, k12.chat.senior, k12.chat.teacher

Subject: Re: KILLING THE CAT CONTEST

In article [deleted] writes:
>I've got this extra cat. And i'd like to do
>some kind of lottery. I'll wait one week to kill the

>cat. Whoever comes up with the best way to snuff the
>life out of this subspecies, i'll do it and report
>the outcome.

This is by far the most vile post I have seen since I
first began reading newsgroups. I don't care if it's
on alt.tasteless, or alt.flame, but please do us a
favor and remove it from other groups that have
nothing whatsoever to do with it
(k12.chat.senior/teacher? Come on!)

Experienced members of the on-line community probably would recognize such postings for what they are. Nonetheless, they may trap the unsuspecting and for this reason are sometimes referred to as "newbie bait," or more whimsically, "trolling for newbies" (Merchant, 1994).

Unlike this relatively benign form of baiting, however, probably motivated by a juvenile desire to derive pleasure from the confusion of others, the extremely personal and often malicious nature of flaming allows the original sender to increase his reputation at the expense of another. But the flame often draws a reflexive response, motivated by a desire to publicly defend one's reputation, which may be even more insulting or offensive than the original. The statement-response cycle quickly escalates, often drawing into the battle those "lurking" (users who simply read the postings and do not post their own messages) on the sidelines, and spreading the war. Libelous flames often can result.

B. Dust off reform proposals

At the center of many defamation reform proposals is the notion that the issue of the truth or falsity of the defamatory statement is critical to a determination of reputational harm. As noted earlier, since the *Times* ruling, the courts have subjugated the issue of truth or falsity to the defendant's motivation in publishing (actual malice or recklessness).

The two key reform proposals are the Uniform Correction or Clarification of Defamation Act and the Annenberg Libel Reform Act. The main purpose of the Correction Act, adopted by the American Bar Association in 1994, is to resolve defamation disputes before they reach court by encouraging both plaintiffs and defendants to consider a timely and adequate correction or clarification. The plaintiff, thus, has the opportunity to vindicate his or her name expeditiously, and the defendant has a way to avoid costly litigation.

In some respects, the act is an improvement over current defamation law. Argues Dragas. "[It] shifts the focus away from fault and the defendant's state of mind (i.e., the actual malice or negligence standards) to the falsity of the statements;...

and it applies to all defamations, public or private, media or nonmedia, thus simplifying the process of gaining redress for a defamation action" (Dragas, 1995, p. 144).

On the other hand, the act has several shortcomings, both from the plaintiff's perspective and from that of the defendant. The act's most significant concern for plaintiffs is that it "gives defendants a substantial window of opportunity for issuing a clarification or correction, while still prohibiting the plaintiff from suing for punitive or general damages" (Dragas, 1995, p. 145). For media defendants, the major concern is that "there will be intense pressure to publish a 'correction' quickly in order to abort a potential libel suit," and that may "encourage hasty judgments about the statement's accuracy and, in the process sacrifice a reporter's reputation and the media's credibility on the altar of expediency" (Dragas, 1995, pp. 146–146).

The second proposal, the Annenberg Libel Reform Act, offers a more comprehensive approach to handling defamation disputes, one whose "ultimate purpose is the timely dissemination of truth" (Libel Reform Act Proposal, 1988, p. 10). The proposed act focuses on retractions, plaintiff replies and declaratory judgments as remedies in defamation actions, rather than the award of monetary damages. The act outlines a three-stage process for resolving the dispute:

Stage I requires potential plaintiffs to seek a retraction or an opportunity to reply from the defendant before filing a lawsuit. Such requests must be made within 30 days of the publication date; if the plaintiff fails to do so, he or she may not pursue further legal action against the defendant. Neither may the plaintiff pursue a lawsuit if the defendant complies with the request to publish a retraction or a reply. The act defines a *retraction* as "a good faith publication of the facts, withdrawing and repudiating the prior defamatory statements," and a *reply* as "the publication of the plaintiff's statement of the facts" (Libel Reform Act, §§ 3(b) and 3(c), 1988). The publication of a retraction or a reply must be made within 30 days of the request.

Stage II, the declaratory judgment option, comes into play if the defendant refuses to grant a retraction or right of reply. Either party may seek a declaratory judgment trial, but the only issue litigated is the truth or falsity of the defamatory statement. In other words, the plaintiff won't be able to recover monetary damages and the defendant won't be able to rely on First Amendment rules for protection. Essentially, the *New York Times* rule, which requires a showing of fault on the part of the defendant, becomes irrelevant. "The injury is what matters. The defendant's knowledge, recklessness, negligence or malice are not issues. The only question to be decided is whether the statement at issue was true or false" (Libel Reform Act Proposal, 1988, p. 11). The plaintiff bears the burden of showing falsity by clear and convincing evidence, and the loser must pay the winning attorneys' fees.

Stage III takes effect when neither side selects the declaratory judgment option and is unwilling to settle without a trial. During this stage, the plaintiff may begin a defamation action and seek monetary damages, though recovery is limited to actual damages.

The proposed act offers several other reforms: It eliminates the distinction between media and nonmedia defendants and between libel and slander, basing all damages on actual injury. It curtails the use of alternative causes of action, such as invasion of privacy and infliction of emotional distress, often used by lawyers to evade the rules of libel trials. It seeks to clarify the distinction between fact and opinion, increasing protection for editorials, letters to the editor, and satire. It also broadens the neutral reportage privilege, protecting defendants who quoted accurately the defamatory statements of identified sources, provided the statements involved a matter of public concern (Libel Reform Act Proposal, 1988, pp. 12–13).

The Libel Reform Act goes a long way toward refocusing defamation law on its historically central purpose: to protect individuals from reputational harm. Specifically, the act's philosophical framework—built on the notion that the critical issue to be determined is the truth or falsity of the defamatory statement—is encouraging.

The act also brings some concerns, however. First, the 30 days that a plaintiff has from the date of publication of the defamatory statement in which to request a retraction or a reply is troublesome. If the plaintiff does not become aware of the defamation until after the 30 days, and misses the retraction-or-reply deadline, he or she accedes the right to bring a future cause of action. Second, by establishing negligence as the minimum level of fault that a plaintiff must prove in a Stage III lawsuit, the act recognizes that it may not amend the constitutional requirement of a higher standard (actual malice) for public officials and public figures. If a defamation action makes it to this stage, therefore, the court is forced to turn its attention to the befuddling distinctions between public and private people and public and private issues and away from a consideration of truth or falsity of the defamatory statement.

Could the Annenberg proposal be applied to on-line communications? One of its underlying concepts—that a plaintiff should have a right to reply to the defamatory falsehood—certainly seems in tune with the reply functions already inherent in computer-based bulletin boards and electronic mail systems. As noted earlier, users of such technologies have the opportunity to respond to flames by posting their own responses. In addition, the libel reform proposal offers, through its declaratory judgment stage, a judicial determination of the truth or falsity of the statement in question. This remedy is key, both off and on line, since most plaintiffs seek not only the opportunity to respond to the defamation but also to have their names cleared through a declaration of the truth.

CONCLUSIONS

Defamation law, as it has evolved since the *New York Times* decision was handed down in 1964, has been so broadly applied and with such conflicting brushstrokes that plaintiffs have little legal recourse when the mass media sully their reputations. Judicial application of the publickness standards enunciated in the *Times* decision and its progeny has resulted in confusing adhoc balancing that gives an almost absolute blanket of protection to the media and rarely regards the citizen's right to be free from reputational harm.

While it is important that the courts be vigilant in assuring a free press, this chapter suggests that the courts have been less vigilant in assuring an individual's right to his or her good name. The age of cybercommunications presents a propitious time for the courts to rebalance the scales of justice to take into account the rights of the individual. Efforts to reform defamation law need to be dusted off, revisited and revised to acknowledge the new ways people communicate with each other and the new ways they go about developing citizenship.

The traditional mass media are slow to embrace the potential of the emerging technologies, and as a result, no court has rendered a judgment in a case brought against a traditional media outlet for on-line defamation. But, the increasing use of computer-based communications technologies makes inevitable a realization that the law's emphasis on the public status of the plaintiff and the public nature of the speech is irrelevant. What remains relevant in the on-line world, though it has been slighted in the physical world, is the law's original intent: to protect individuals from reputational harm suffered by the publication of false statements of fact. An emphasis on those two issues—reputational harm and the truth or falsity of the defamatory statement—will bring timely and necessary change to the law of defamation.

REFERENCES

Anderson, D. (1984). Reputation, compensation, and proof. *William and Mary Law Review 25*, 747–778.

Associated Press v. Walker, 388 U.S. 130 (1967).

Barron, J. A. (1967). Access to the press: A new First Amendment right. *Harvard Law Review 80*, 1641–1670.

Barron, J. A. (1973). *Freedom of the press for whom?* Bloomington, IN: Indiana University Press.

Bartlett, K.T., & Kennedy, R. (Eds.). (1991). *Feminist legal theory: Readings in law and gender.* Boulder, CO: Westview Press.

Bellah, R. N. (1986). The meaning of reputation in American society. *California Law Review 74*(3), 743–751.

Bezanson, R., Cranberg, G., & Soloski, J. (1987). *Libel law and the press.* New York: Free Press.

Blanchard, M. (1982). Filling in the void: Speech and press in state courts prior to Gitlow. In B. Chamberlin & C. Brown (Eds.), *The First Amendment reconsidered: New perspectives on the meaning of freedom of speech and press* (pp. 14–59). New York: Longman.

Borden, D. L. (1988). *The invisible plaintiff: Protecting the rights of private individuals in the wake of Hustler v. Falwell.* Unpublished master's thesis, Stanford University.

Borden, D. L. (1997). Patterns of harm: An analysis of gender and defamation. *Communication Law and Policy, 2*(1), 105–141.

Carter, B., Franklin, M. A., & Wright, J. B. (1985). *The First Amendment and the Fourth Estate* (3rd ed.). Mineola, NY: Foundation Press Inc.

Cubby, Inc.v. CompuServe, Inc., 776 F. Supp. 135 (S.D.N.Y. 1991).

Curtis Publishing Co. v. Butts, 388 U.S. 130 (1967).

Dragas, M. L. (1995). Curing a bad reputation: Reforming defamation law. *University of Hawaii Law Review, 17*, 113–164.

Dun & Bradstreet, Inc. v. Greenmoss Builders, Inc., 472 U.S. 749 (1985).

Franklin, M. A. (1980). Winners and losers and why: A study of defamation litigation. *American Bar Foundation Research Journal*, 455–500.

Gertz v. Robert Welch, Inc., 418 U.S. 323 (1974).

Hutchinson v. Proxmire, 443 U.S. 111 (1979).

Keeton, W. P., Dobbs, D. B., Keeton, R. E., & Owen, D. G. (1984). *Prosser and Keeton on torts* (5th ed.). St. Paul, MN.: West Publishing Co.

Lang, G. E., & Lang, K. (1990). *Etched in memory: The building and survival of artistic reputation.* Chapel Hill, NC: University of North Carolina Press.

Levy, S. (1995, December 25). The year of the Internet. *Newsweek*, 21–30.

Libel Reform Act (1988). *Proposal for the reform of libel law: The report of the libel reform project.* Washington, DC: Annenberg Washington Program.

MacKinnon, R. C. (1995). Searching for the Leviathan in Usenet. In S. G. Jones (Ed.). *Cybersociety: Computer-mediated communication and community* (pp. 112–137). Thousand Oaks, CA: Sage Publications.

Medphone Corp. v. Denigris, No. 92-CV-3785 (D.N.J. filed Sept. 11, 1992).

Meiklejohn, A. (1960). *Political freedom: The constitutional powers of the people.* New York: Harper & Row, 1960; reprint (1979), Westport, CT: Greenwood Press.

Merchant, J. (1994). *Ethnographic description of the 'Net culture.* Unpublished undergraduate paper, Philadelphia:Temple University.

New York Times v. Sullivan, 376 U.S. 254 (1964).

Norton, M. B. (1987). Gender and defamation in seventeenth-century Maryland. *William and Mary Quarterly*, (3rd ser.). *44*, 3–39.

Philadelphia Newspapers, Inc. v. Hepps, 475 U.S. 767 (1986).

Pollock, Sir William (1894). *On torts.* London: F.H. Thomas Law Book Co.

Post, R. C. (1986). The social foundations of defamation law: Reputation and the Constitution. *California Law Review, 74*(3), 691–741.

Probert, W. (1962). Defamation, A camouflage of psychic interests: The beginnings of a behavioral analysis. *Vanderbilt Law Review, 15*, 1173–1201.

Rindos v. Hardwick, Supreme Court of Western Australia, unreported judgment, No. 940164 (1994).

Rosenblatt v. Baer, 383 U.S. 75 (1966).

Skolnick, J. H (1986). Foreword: The sociological tort of defamation. *California Law Review, 74*(3), 677–689.

Smolla, R. A. (1986). *Suing the press: Libel, the media, & power.* New York: Oxford University Press.

Stratton Oakmont Inc. v. Prodigy Serv. Co., 23 Med. L. Rptr. 1794 (N.Y. Sup. Ct. May 24, 1995).

Suarez Corp. Industries v. Brock N. Meeks, No. CV-267513 (Ct. C.P., Cuyahoga Co. filed March 22, 1994).

Tedford, T. L (1993). *Freedom of speech in the United States* (2nd ed.). New York: McGraw Hill, Inc.

Time, Inc. v. Firestone, 424 U.S. 448 (1976).

Time, Inc. v. Hill, 385 U.S. 374 (1967).

Wilson, P. J. (1974). Filcher of good names: An enquiry into anthropology and gossip. *Man* (New Series), IX, 93–102.

Wolston v. Reader's Digest Ass'n, 443 U.S. 157 (1979).

Yang, C. (1995, February 6). Flamed with a lawsuit, *Business Week*, 70–72.

III

REPORTING

The Campus Press:
A Practical Approach to On-Line Newspapers

Bruce Henderson
Jan Fernback
University of Colorado

Most on-line publishers seem to have a "field of dreams" philosophy: Build a World Wide Web newspaper, and readers will come. But building an on-line newspaper is more complicated than building a baseball diamond in an Iowa cornfield.

Since its humble beginnings in Geneva, Switzerland, in 1991 (Tittel & James, 1995), the World Wide Web (WWW) has garnered millions of WWW pages, all vying for readership. Developed by the European Laboratory for Particle Physics (CERN) to enable users to share information using a simple "point and click" system, the WWW allows on-line users to retrieve text, graphics, video, and audio data through an easily navigable browser software package. With the explosive popularity of the WWW among businesses, organizations, and individuals, an on-line publisher must dream beyond simply getting to that first information base in order to attract a loyal following.

A successful on-line newspaper must have excellent, local information; it must be designed well; it must be interactive; and it must successfully build a community around the information and interaction. This chapter is a case study of one such on-line publication.

The Campus Press, a weekly student-produced newspaper at the University of Colorado (CU) at Boulder, circulates 8,000 print copies on the CU campus. It began publishing a weekly WWW version in the fall semester of 1994. Early in the spring semester of 1995, nearly 900 on-line readers of *The Campus Press* viewed an average of 5,000 documents each week. The criteria for building the online *Campus Press* and other successful on-line publications are described below.

INFORMATION, INTERACTION, COMMUNITY

Building an on-line newspaper involves more than shoveling all the stories and press releases available into a WWW publication and hoping people will become regular readers. Publishing an on-line newspaper requires the nurturing and development of information, interaction, and community. These are the same elements necessary for a successful print newspaper. The difference on-line is that these relationships can be quicker and more binding on a larger geographic scale.

Newspapers build readership through content. What binds a newspaper to its readers is information about the readers' community as well as the readers' trust in that information. This information can take the form of local stories, or national and international stories related to the interests of the community; but it can also take the form of reader-contributed material, such as letters to the editor, announcements about cultural events, or, in some newspapers, news about clubs and organizations. This reader-contributed content—which is a type of interaction among readers, the newspaper and others in the community—can also take the form of display advertising and classified advertising, where the reader has almost absolute control over the content. At present, the advertising information makes up about 60% of most newspapers; but with rising newsprint costs, some newspapers are having to cut the amount of news in a paper to as little as 30% of the content (Underwood, 1993).

An on-line newspaper has a virtually unlimited "newshole." Instead of editing stories to fit the available space or condensing the geometry of the actual page layouts (Rosenberg, 1995), editors can revise stories for completeness. On-line stories can be linked electronically to archived, background stories and to analyses. Similarly, ongoing stories can be condensed, depending on how many good, background stories are linked to the breaking news. On-line story summaries become important as a way for busy people to browse quickly through a publication; if the reader wants more information, it's only a click away. Thus, on-line news publications have distinct advantages over hard copy versions. With the online *San Jose Mercury News* or the online *Boston Globe*, for example, users can enter keywords to expedite searches for classified or news information (Geier, 1995). With the *Los Angeles Times' TimesLink*, users can search years' worth of archived articles quickly and conveniently using keywords (Mannes, 1995). In addition, on-line letters to the editor can be published instantly. Readers can send their comments to a common area, called a newsgroup, where the comments can be read by any subscriber. This material, in turn, can be edited and later used with permission in the print version of the newspaper.

These features enhance the reliability and credibility of on-line publications. As some observers (e.g., Stoll, 1995) continue to criticize the Internet for its abundance of unsourced or unreliable information, users will increasingly congregate around information they can trust—such as that found in traditional media sources, including the local newspaper. With the established reputations of local newspapers at stake, users will be likely to rely on their on-line newspaper to provide not only credible local information, but also to find and edit the credible information that is available on the Internet.

While print newspapers are often constrained by the rising costs of newsprint (Rosenberg, 1995) to the spatial detriment of community-oriented content, on-line newspapers have the space they need to devote to building strong ties to the community. For example, community areas could be created within these newspapers where clubs, organizations, or other interest groups could post announcements, full meeting agendas, or other information. Civic and government organizations also could post information on, for example, city council meeting agendas, volunteer opportunities, or recycling centers, that users may not otherwise know how to obtain. Likewise, students in grade schools could become involved with building online pages as a way to bring young readers into the newspaper.

On-line newspapers also could provide live forums for discussion, where people could click into a conversation and participate by typing their comments and watching as they appear instantly onscreen through chat software. The print newspaper could announce topics for discussion related to current events; some of the on-line comments could be edited and used in the print publication. On-line newspapers also could feature special guests in the chat forums. For example, the mayor of a community could be a guest in the chat area, similar to talk radio hosting an important political figure, and on-line readers could ask the questions. Or, members of a band might appear as guests in a chat forum as a way to publicize an upcoming appearance.

It is through this kind of instant interaction—contributed reader content, newsgroups and chat—that an on-line newspaper could build a strong community of readership. The on-line journalism phenomenon has helped to publicly illuminate the fact that most of the content in both print and on-line newspapers is reader-contributed. Letters to the editor, advertisements, calendar events, and press releases by far constitute the bulk of both types of newspapers; thus, the concept of journalism itself may change as more readers recognize that they dictate, in large measure, newspaper content. So, while the usefulness of on-line resources remains questionable for journalists seeking a tool to aid in their critical public role, on-line newspapers themselves do help to foster community building. For the cost of a phone call, out-of-towners could maintain contact with their communities through their hometown newspapers; online public forums within these newspapers could be created to serve as a "virtual" town square; and users could access text, audio, and video data through one medium. Clearly, the fun and ease of using the Web, in concert with its lack of spatial constraints, could allow online newspapers to facilitate community bonding.

The same is true for advertising. Readers could submit ads on line. Many advertisers, in fact, send their material via modem to print newspapers. The WWW could make this process easier by supplying on-line forms. As with news content, there aren't the same space or photo limitations on-line that there are in print, since the FCC has not limited the amount of bandwidth publishers may consume. Perhaps more importantly, on-line ads could be interactive. With a simple click, a reader could ask for more information, send a fax to the advertiser, or order something from an advertiser. Classifieds that offer this interactivity have the potential to surpass the response rate of print classified ads; an on-line reader has only to click on a classified to respond. While complete interactivity is not yet a reality, many of these features are now available through, for example, the *Raleigh News* and *Observer's NandO Net* (Mannes, 1995).

It's important to note that the ability to send an advertiser a fax from the WWW means the advertiser doesn't need Internet access to get responses from an on-line ad. Many major advertisers, such as General Motors and Microsoft, already have built WWW pages. They are seeking popular WWW sites and are willing to pay for a link from those sites to their advertising. For local on-line newspapers, providing this link would be similar to publishing national advertising. But local advertising has always supplied the bulk of revenue for newspapers. A local newspaper could offer a local advertiser server space and assistance building ads on line. A local on-line newspaper also could offer print/on-line combination opportunities. For example, a print ad might refer the reader to an on-line catalog. Ultimately, an on-line newspaper might better serve a print newspaper community by offering more news, more opportunities for readers to contribute and participate, and more ties between readers and advertisers.

WHY PUBLISH ON THE WORLD WIDE WEB?

Although computer-mediated communication has allowed general users to connect to each other and to vast storehouses of information for more than a decade, many people have been deterred by the complexity and amorphousness of navigating the Internet through Unix commands. The WWW has all but eliminated this reluctance. For example, shortly before a University of Colorado football game in the fall of 1994, a group of 150 School of Journalism and Mass Communications alumni gathered in a computer lab for a demonstration of the online *Campus Press*. Few had ever been on the Internet. Some had never tried computers. After a brief presentation, everyone was smiling and surfing through the newspaper. Soon they were taking virtual Internet journeys to different parts of the globe. Why should newspapers be concerned about this kind of transformation? Here are a few of the reasons:

- Building WWW pages is easy. The University of Colorado had offered students, through their free university e-mail accounts, a chance to build their own Web pages in the spring of 1995. By summer, there were 1,016 registered pages. Commercial Internet providers, such as America Online, are now offering their customers the opportunity to build their own personal pages for free.

- The number of commercial newspapers that publish on the WWW has grown at a phenomenal rate; approximately two new newspapers per day come on-line, according to Steve Outing, of *Planetary News* and *Editor & Publisher*. Outing tracks on-line newspapers and publishes a WWW site called Media Info Interactive. In the time since the introduction of Mosaic in 1993 (Sussman & Pollack, 1995), a free Internet browser application that allows the display of headlines, photos, text, sound, and movies in a format that simulates a publication, the number of commercial newspapers publishing on the WWW skyrocketed from zero to 424. Of those 424 newspapers, 326 are U.S. newspapers (Outing, 1995a). Most major newspapers in large markets are now on line; for example, *The New York Times*, the *Philadelphia Inquirer*, the *Los Angeles Times*, the *Boston Globe*, the *Houston Chronicle*, the *Miami Herald*, the *Denver Post*, and both the *Chicago Sun Times* and the *Chicago Tribune*.

- Access to the WWW over the Internet is cheap and getting cheaper. Universities provide free, unlimited accounts to students and faculty. Commercial Internet providers charge from $5 a month plus a $2-an-hour access charge, to $35 a month for unlimited access. As of the summer of 1995, businesses could put their information on the WWW for as little as $50 a month. When telephone and cable companies start offering Internet access, rates are expected to decline as competition sharpens (Gilder, 1995). Presently, AT&T is working to develop new Internet access technologies, such as symmetrical digital subscriber loop (SDSL) and cable modems. Bell Atlantic is already testing SDSL technology in homes for approximately $10 per month. However, Internet Service Providers (ISPs) are developing browsers and other Internet software platforms that are compatible with any operating system, so competition between them is indirect. Although ISPs depend on telephone companies and potentially even cable companies for connections, while telcos and cable companies depend on ISPs to supply Internet services, provide new customers, and for technical support, pricing competition is expected to intensify (Gilder, 1995).

Why should newspapers explore publishing on line?

- Fear. What if an entrepreneur enters the market and captures a big chunk of classified advertising on-line? It could happen. Producing an on-line version of a local newspaper could help the newspaper retain, if not increase, its advertising market share. Moreover, producing both a print and on-line newspaper could create synergistic benefits for both. Each could be used to generate visibility for the other; each could trade content, especially content generated by on-line readers. For example, the *Fayetteville* (North Carolina) *Observer-Times* introduced its Web version of the paper in its print version. Using a print ad featuring computer mice driving on a road with accompanying copy reading, "New Highway Connects Fayetteville to the World," the *Observer-Times* touted its new electronic version. The paper also conducted a "community meeting" to introduce the service and to teach citizens how to use the Internet (Outing, 1995b). Also, on-line copyright issues become less problematic when print and on-line versions of the same newspaper share copyrighted content. Because new copyright issues are still being debated within legal circles, pre-existing copyrights that are legally recognized in print probably will subsequently be recognized in the on-line environment.

- It's cheap. Many commercial Internet providers are hungry for good content and eager to work with newspapers on pricing. There are no printing press costs, no circulation costs.
- There is nearly unlimited space. Plus, stories can be linked to analyses, or to related archived material.
- News can be more immediate on-line than in print.
- Print, video, and sound can be used to enhance the on-line newspaper.
- There are more opportunities for entertainment, such as interactive games.
- Celebrities can log in for interviews conducted by readers.
- Elementary school children are learning all about computers and the Internet. To them, it's as easy as using a VCR. When they reach prime newspaper readership age, they are likely to continue their preferences for interacting with sophisticated new media rather than with traditional newspapers (Lapham, 1995).

Of course, it makes no sense for a local newspaper to also publish an online WWW edition in an area where there is little Internet access. Thus, it is important for a newspaper to learn about the number of Internet users in its circulation area; Internet use could be probed in every newspaper readership survey conducted.

Industry estimates (Calem, 1992) indicate that about 30% of U.S. households have computers; about one third of these computers have modems, and about one third of these computers are used to access commercial on-line services or the Internet. (Soon, having access to a commercial service will mean having Internet access; all of the commercial services are shifting to offering WWW access.) However, with access to on-line services in the work environment and increasingly in public areas such as kiosks, libraries, and even cyber cafes, the gap that distinguishes users from those with no access is rapidly closing.

The number of Internet providers in an area also can be an indication of access. Some providers will talk openly about how many users they have. In Boulder, for the *Campus Press*, it was easy to decide to put out an on-line edition. The University of Colorado provides about 32,000 free Internet accounts for students and faculty; there are 38 computer labs tied to the Internet throughout the university and in dorms; and the Boulder area has several government labs that are tied into the Internet (the National Center for Atmospheric Research, the National Renewable Energy Laboratory, and the National Institute of Standards and Technology). There were virtually no costs involved, except for $300 per semester to pay a student editor. The newspaper is student produced, and there was no cost to the newspaper for 24-hour access to an Internet server.

ON-LINE CAMPUS PRESS READERSHIP

A counter keeps track of the number of "hits" on items that are read in the on-line *Campus Press*. It's like having instant marketing data. Using a statistical analysis program called Getstats, an on-line readership survey of the *Campus Press* was conducted between February 24 and March 30, 1995. The survey indicated that the popularity of sections in the on-line *Campus Press* are, in order, fun (which includes entertainment stories from the print version, plus many entertainment items from the Internet), the coffee house (which is a chat area where readers can "talk" to each other), news (which contains news stories from the print version, as well as related Internet stories), sports (which includes stories from the print version, plus related Internet links), community (which is an area for campus clubs that is not duplicated in the print version), internet (which includes Internet search resources), calendar (which includes the calendar from the print version), services (advertising), opinion (editorials and letters to the editor, plus feedback from on-line readers), weather (which includes up-to-the-minute National Weather Service data, as well as ski reports), and mail (which is a way for readers to access their e-mail accounts).

Generally, readership is lowest between 3 a.m. and 5 a.m., peaks at 11 a.m., declines slightly over the lunch hour, picks up again between 1 p.m. and 2 p.m., then drops sharply between 4 p.m. and 5 p.m. Generally, the largest readership is on campus computers, then high-tech businesses and a commercial on-line service. There also is a small but consistent national and international readership.

What do these numbers mean? Many people access the *Campus Press* at work, with the highest readership between 8 a.m. and 5 p.m. Many of the readers are off-campus. Most of the readers are seeking entertainment and communication with each other. News and sports are important components. The readership is largely local. A look at comments sent by readers reveals a little more information. One reader, who stated he was from Oslo, Norway, said he is a CU grad who enjoys following CU sports on-line. Another reader, an alumna living in Georgia, said she enjoys following the news about CU. Another reader, from Chicago, said he enjoys reading his son's column on line because he doesn't get the print version. A reader living in the southeast said that she'd like to meet other CU alumni living in her area. Students and others from around the country have asked for housing or other information about Boulder.

This kind of information is similar to the information newspapers obtain through expensive readership surveys. But the computer can generate the numbers instantly, accurately and for free with the use of statistics programs available from the WWW. If an advertiser wants to know whether an ad has been read, and some basic information about who read it, the computer can churn out the numbers in seconds. On-line newspapers, just as print newspapers do, can use this marketing data to help build a larger readership.

THE FUTURE

The WWW, since the formal introduction of Mosaic in 1993, has already caused a revolution in on-line publication. But the revolution is getting a bit stale as the number of new WWW information sites grows. As the amount of information expands, the credibility of the Internet tends to decline (Stoll, 1995). There is too much information of little, no or questionable value. This represents the coming decline in interest in the Internet; it also represents the opportunity for newspapers, which can supply credible information. Newspapers need to apply the same careful reporting and editing techniques that they always have when producing an on-line publication.

They also need to demonstrate the same kind of savvy in knowing their readership, be able to change and grow according to the on-line readership demands, yet keep the priority focused on content. Similarly, newspapers need to design on-line publications that go beyond the on-line designs produced by computer technologists. Many print newspapers, when they first converted to powerful personal computers and pagination, made the mistake of producing gaudy, cartoon-colored pages because it was technologically possible. The same thing is occurring now on the WWW. On-line publications need to be designed not by technologists, but by people who have a good understanding of news presentation and design and who are not designing pages simply because the technology makes it possible. Advertising opportunities need to be more well-developed in new publications. On the one hand, large-scale advertisers are creating their own Web sites. On the other, local advertisers are doing very little. And if they are, they are more than likely doing it through their local Internet provider, rather than through the local newspaper.

Finally, newspapers need to build alliances with television stations. (Editor's Note: See chapter 5 for a case study on just this topic.) Currently, there is a trend for television stations to create WWW sites. As on-line video becomes the next World Wide Web trend, newspapers and television need to work together. This venture is a reality at the University of Colorado; journalism students in both print and broadcast produce full news stories accompanied by related video reports as part of the on-line *Campus Press*. Print and broadcast students, in concert, produce audio and video clips within the *Campus Press* on-line environment; soon at the University of Colorado, on-line publication will evolve into multimedia production. The priority will be on news information, interaction and community. With that formula, perhaps more readers will come—and will return often.

REFERENCES

Calem, R. (1992, December 6). The network of all networks. *The New York Times*, p. 12f.

Geier, T. (1995, November 20). Tracking the news in cyberspace. *U.S. News & World Report, 119* (20), 106.

Gilder, G. (1995, December 4). Telecosm: Angst and awe on the internet. *Forbes*, 112f.

Lapham, C. (1995, July 1). The evolution of the newspaper of the future. *CMC Magazine, 7* (http://sunsite.unc.edu/cmc/mag/1995/jul/lapham.html).

Mannes, G. (1995, November). The new news "paper": Electronic editions of newspapers. *Popular Mechanics, 172* (11), 66.

Outing, S. (1995a, December). *Editor & Publisher online* (http://www.mediainfo.com/edpub).

Outing, S. (1995b, October 31). Marketing tips for an online newspaper service. Stop the Presses? *Newspaper New Media News and Analysis* (http://www.mediainfo.com/edpub/ep/stop1031.htm).

Rosenberg, J. (1995, December 2). Trimming the broadsheet: Does printing on 50 newsprint rolls mean busting the SAU and shrinking the news hole? Different approaches at three dailies. *Editor & Publisher Magazine*.

Stoll, C. (1995). *Silicon snake oil: Second thoughts on the information highway.* New York: Doubleday.

Sussman, V., & Pollack, K. (1995, November 13). Gold rush in cyberspace. *U.S. News & World Report, 119* (19), 72f.

Tittel, E., & James, S. (1995). *HTML for dummies.* Foster City, CA: IDG Books.

Underwood, D. (1993). *When MBAs rule the newsroom.* New York: Columbia University Press.

Technology and Journalism in the Electronic Newsroom

L. Carol Christopher
University of California, San Diego

Technological change takes place in both a social and a historical context, which exist both internal and external to the organization; each context is embedded with interests, beliefs, and values about social status, the nature of meaningful work, and what problems technology can or should solve (Thomas, 1994, p. 4). Technology makes journalistic work more abstract in numerous ways and thus changes the nature of journalistic skill: The firsthand sentience—the sounds and sights and smells of the production of the printed word and, increasingly, human interaction as a central facet of journalists' work—are increasingly displaced by computerized interactions and electronic texts. Technology both serves as a means of rationalizing journalistic work in a way that removes, erodes, or creates opportunities for communicative and coordinative activities in the newsroom, and presents opportunities for extending old skills or creating new ones.

Increasingly, communication theorists have begun to focus on the socially constructed nature of reality, and to examine the workplace as a potential site of conflict over different interests in decisions regarding products, our culture, and our environment (see Deetz & Mumby, 1990). Computer technologies are not neutral: They have inherent characteristics that constrain their potential in the workplace, since much of what has driven the design, development, and deployment of workplace technology is that it can be controlled more easily, and is more subject to rational principles, than humans. Machines do not tire or object to repetition or rationalized activities as do humans, and are more consistently capable of precision, thus rendering production processes more predictable, and creating a more secure environment for capital investment and profit.

Newspapers are a social construction of reality—a representation of what's important to know, and by extension, what isn't—which occurs as journalists jointly participate in the selection, framing, and presentation of events and ideas which then become The News. But this construction is also influenced by decisions and processes external to the newsroom—including decisions about labor, about profit, about technology. This chapter explores three different implementations of technology in newsrooms: electronic editing, pagination, and computer-assisted reporting, as a means for examining the ways in which newspapers are using newsroom technology to gain greater control over both the content and appearance of the newspaper, but also to gain greater control over the labor process of workers in the interest of the accumulation of capital. It concludes with an examination of new forms of work organization in the newsroom. In doing so, it links considerations of the social construction of reality by major media corporations with questions raised by labor process analysts, especially as raised by Braverman, Shaiken, and Zuboff. Labor process analysts challenge the social agreements between management and labor, which they see as resting on yet another social construction of reality, a set of practices, a discourse, a historical process: They question which work, rights, and powers accrue to owners/managers, which to labor. Among their concerns is that while the technological demands of work require greater education, training, intelligence, and mental effort, workers and society find conditions of industrial and office labor increasingly unstable, unsatisfactory, and characterized by mindlessness, bureaucratization, and alienation.

Braverman (1974) for example, challenges accounts of the organization of modern industry by modern social science to "accept all that is real as necessary, all that exists as inevitable, and thus the present mode of production as eternal" by providing a concrete and historically specific analysis of the relationship between technology and machinery on the one hand, and social relations on the other (pp. 16–17). Shaiken (1986) argues that technological development is a consequence of choice, both social and economic, and that the workplace is shaped by the interaction of workers, managers, and technology (p. 13). Management, whose goals shape the design and implementation of technology in the workplace, seeks to control production as well as the activities of workers. Technology, and particularly information technology, may be designed in a way that leads to reduced skills and levels of worker input and decision making, as well as tighter control over work and workers' activities. The result is boring, stressful work.

While managers seek higher productivity from such a system, its actual costs may be lowered quality and productivity, since a system that reduces human input cannot benefit from human skill, talent, experience or creativity. In the newsroom, it isn't just journalists who lose: Newspapers lose the synergy that comes from creative and sometimes casual human interactions. The effects of management decisions, however, may be mediated by a broad range of factors, such as technology design, as well as the presence and strategies of organized labor.

Zuboff (1988) found that computerized work is more mediated than older technologies—workers must deal with an increasingly symbolic, or abstract, world, relying more heavily on their computers and less on their own sentient experience of the world (p. 75). In other words, working with computers offers a new way of knowing the world. Depending on the choices made about the deployment of information technology, workers may gain communicative and coordinating skills and intellectual competence, or they may be pushed "further into the sentient but mute terrain of fatigue and nervous exhaustion" (p. 123).

EDITORS IN THE ELECTRONIC NEWSROOM

Since the early 1970s, newsrooms have seen the introduction of video display terminals (VDTs), pagination, libraries, on-line and database research, remote transmission and delivery, photo desks (enhanced color capabilities and digital photo transmission and storage), and photo editing capabilities. Many newspapers, in meeting technical requirements of computer systems, have also remodeled their newsrooms so that they, in the words of more than one journalist, "look like an insurance office" more than the newsrooms of the recent past.

The capital investment decisions that result in the arrival of shiny new newsroom technology are made in complex environments. As Thomas (1994) contends, we must "conceive of the relationship between technology and organization as mediated by the exercise of power, that is, by a system of authority and domination that asserts the primacy of one understanding of the physical world, one prescription for social organization, over others" (p. 5). Corporate owners and managers see in these technologies a means for lowering labor costs, expanding old markets and reaching new ones—in short, a means for increasing profits. The technological choices a manager makes represent her or him symbolically within the organization; the selection of a particular technology may be a way of gaining status or influence within the organization, and influencing or sabotaging its goals (p. 6). In a 1994 speech, George R. Cashau, former senior vice president/technology for the Newspaper Association of America, stated that to be useful to newspapers, technology must, in addition to positioning newspapers as key information providers, produce revenue, be faster, more efficient, economical, and easy to use. By way of contrast, George Gerbner (1995) writes that "the basic problem of journalism is... media conglomeration and the consequent reduction of staff, diversity, and *time* to do an adequate job. For journalists, [newspaper technology] means further loss of control to a few wholesalers and global marketers of media 'software'" (p. 5). Owners, managers, journalists, and technicians have different interactions with these technologies, and the technology leaves different marks on the work of each of these groups. Thus, for each group the technology has a different meaning.

Electronic editing

Electronic editing was the first step toward a fully computer-paginated newspaper, one in which entire pages are typeset, rather than typesetting each story, headline, and caption individually and then pasting them up in the composing room, or backshop. Pagination reduces the amount of work that must be done in the backshop, and consequently the number of employees in that area. The editor's function expanded to include the remaining tasks—responsibilities such as typesetting, printing, and proofreading—which had formerly resided in the backshop (Kurtz, 1980, p. 55). Not only were there more tasks, but they were tasks that demanded the skill to accomplish them in the abstract world of the computer, rather than the concrete activity of the backshop (Zuboff, 1988, p. 126). Despite greater newsroom control over copy flow and improvements in mechanical accuracy, there were costs in other areas. For example, wire copy (i.e., stories transmitted by news services, such as the Associated Press, United Press International, or Reuters) began to arrive in the newsroom at 1,050 words per minute around 1976 (compared to 66 words per minute by teletype, the prior technology). By the late 1980s, wire editors who had previously received hard-copy printouts or had to feed paper tape through a reader before seeing a story, were now receiving copy electronically at nearly 10 times the speed at which it had been sent with the arrival of VDTs, roughly 10 years before. And with the increased transmission speed came an increase in the number of transmissions. Transmission speeds get faster and volume continues to increase as the technological capacity for sending, receiving, storing, and routing transmissions becomes more sophisticated and relatively more affordable: Editors in one study reported they were spending 97% of their time just pushing the buttons that released copy to other editors (Lindley, 1988, p. 486). In the meantime, photos were arriving 10 times faster than previously. And editors disagree over whether typographical errors were harder or easier to spot on the computer screen than on paper copy. Speed sometimes came at the cost of good editing (p. 487).

This early experience with technology brought with it other concerns as well: quality of work life, editor reactions to the technology outside of actual job performance, and whether the demand for technological competence infringed on journalistic quality. Many copy editors rejected the idea that they had been turned into technicians, and strongly disclaimed the notion that VDTs had depersonalized copy editing. Others, however, found the opposite to be true—finding the work depersonalized and isolating, at times producing a newspaper of lower quality. Many missed the support of the now mostly eliminated proofreaders. There was wide disagreement over whether VDT editing required higher concentration or skill than paper editing, suggesting a demand for higher technological skill, but no increase in editing skill requirements.[1] Most editors felt that newspapers had be-

---

1. Although the notion of skill as quantifiable is debatable, I have used Lindley's terms, which suggest that he sees skills as quantifiable, to represent his findings. Newspapers sometimes offer "skills testing" before hiring an editor in order to measure/quantify/evaluate those skills.

come more sophisticated, but that the increased sophistication related more to production process than content. They also expressed concern over the new potential for top-rank editors to read over the shoulders of reporters, city editors, wire editors, and copy editors (Lindley, 1988, p. 489), and about the effect of electronic editing on the quality of their work. Just as the technology brought with it an increased volume of wire copy at an increased speed, for many editors, VDTs and the higher number of tasks that came with them meant that there was less time for the more traditional tasks they saw as their primary journalistic responsibility. The demand for technological skill also took time away from editing. There was an awareness that tension was building between the concurrent demands for technological and traditional skills.

Editors who had worked through the transition from pencil editing to electronic editing were most likely to be aware of these and other effects. Many of these older employees have either reached retirement age or been "bought out" during downsizing and merger scenarios. Findings of future research will reflect the different experience of the greater number of workers who have been working with computer technology at least through college or the beginning of their journalism career. These newsroom workers are more likely to accept a workplace permeated by technology without question compared to earlier journalists—to them it is "natural," not necessarily an issue arising out of changes in economic control or ownership patterns, and subsequent choices that have led to shifting definitions of journalistic skill and quality.[2] Both the constraints and liberating qualities built into existing technologies by a steady historical stream of social choices will influence the way new generations of journalists see their own skills, their ability to control the products of their labor, and the nature of relations of power in their workplace.

Information technology in the newsroom has had, in some ways, a reintegrative quality in the newspaper—reintegrating, in one job, tasks that had been rationalized, or specialized into many separate jobs. But the new technology brought not only new ways of doing old things, but also the capability for new ways of doing new things. Newsroom managers had to consider new ways of organizing newsroom work to accommodate the changes. The newsroom transformation showed up in a number of areas. Among them was the creation of newsroom positions to

---

2. See Zuboff, p. 13: "The most treacherous enemy of such research is what philosophers call 'the natural attitude,' our capacity to live daily life in a way that takes for granted the objects and activities that surround us. Even when we encounter new objects in our environment, our tendency is to experience them in terms of categories and qualities with which we are already familiar. The natural attitude allows us to assume and predict a great many things about each other's behavior without first establishing premises at the outset of every interaction. The natural attitude can also stand in the way of awareness, for ordinary experience has to be made extraordinary in order to become accessible to reflection. Awareness requires a rupture with the world we take for granted...."

help traditional journalists manage technological change: Computer system editors/managers, usually with previous newsroom experience, provided training, advice and troubleshooting for the newsroom. The new technology provided them, and by extension, newsroom managers, with the information necessary for quantitatively assessing the performance of newsroom workers. The flow of work in the newsroom could be surveyed abstractly and symbolically. The skill to interpret that information was most often held by supervising or systems editors and technicians. It was used most frequently for settling deadline disputes with the composing room. The locus of decision making was still management-level editors, and its structure remained hierarchical.

These managers, who often had worked in several newsrooms, became increasingly accustomed to seeing and including large line item numbers on their budgets for technology as it became a more familiar feature of virtually every U.S. newsroom,[3] as they became accustomed to greater control over the production process and demanded ever greater speed, and as newspapers gained an increasing awareness of the added value that strategic combinations of information, technology, and management can bring. One early example was the ability to put out more editions with broader geographical dispersion than before (e.g., Diamond, 1993, p. 235; Kurtz, 1980, p. 57). So-called "zoned editions," with advertising and news tailored to discrete geographical areas, followed not long after.

Pagination

New data-compression technologies, plus cheaper disk storage media, have helped fuel the move toward pagination for newsrooms. But another technological shift also has contributed to the increased presence of pagination systems—the arrival of desktop publishing technology in the newsroom. With the graphical user interface, work that once involved a fairly high level of craft or systems competence can now be produced by novice users in minutes. Further, as this interface has become the primary platform for newspaper graphics, a service industry has grown to provide those graphics—either standard or customized—via dialup or automatic telecommunications transmissions. Work once done in the backshop or later, the newsroom, is often now produced by a service bureau, which distributes material for newspapers across the country and across the globe. The choices of technology for producing camera-ready materials—such as ads, typeset stories, design elements, or whole pages, which are then used to produce printing plates for the presses—for both the newsroom and the advertising department have

---

3. Here are projected capital outlays for 1993–1994 compared to actual 1994 outlays for newsroom technologies nationwide: Editorial computers: projected, $58,622,386 versus 1993 actual, $49,370,951; Electronic library systems: projected, $5,394,787 versus 1993 actual, $2,322,147; Digital darkroom, editorial photo department, electronic picture desk: projected $12,294,799 versus 1993 actual, $9,248,994. Source: NAA Tech News, June 1984. p. 3.

broadened considerably. Where the decision to buy a newsroom system was once mostly one of selecting which vendor from whom to buy, increasingly it involves a choice between proprietary or off-the-shelf systems, and often requires the ability to integrate the two—always with an eye open to the need for support and new developments that might render current choices obsolete.

Often the people who make decisions about the purchase and deployment of technology come from very diverse backgrounds, ranging from data processing to typographical production (composing) to journalism, which all offer very different perspectives on how and toward what ends work should be organized. Often, there is no lost love, or respect, among these different groups. Obviously, this kind of technology has a direct impact on work organization. It provides newsrooms with the much desired ability to maintain control over their work flow. It also makes it possible to evaluate in new ways where the bottlenecks in newsroom production existed.

The decisions that drive the choice of pagination technology are usually based on recommendations from information management, managers whose background is primarily technical, in conjunction with recommendations from newsroom systems managers or mid-level managers, such as assistant managing editors. Discussions of implementation strategy are usually limited to how quickly training could be accomplished, and how much knowledge different types of workers in the newsroom need. Discussion about loss of traditional skill, job enrichment, or increased cognitive effort rarely evolve, most likely for two reasons. First, the worlds of newspaper management versus critical discussions or academic research rarely converge. When they do, they represent a collision of purpose: The broad social context in which social scientists situate the problems which newspaper managers must solve to keep their jobs and satisfy corporate shareholders provides little in the way of immediate relief to these problems. Second, the success of these managers is measured by traditional goals and understandings of efficiency and cost savings, as well as quality, rather than by the standards of a highest common good which social scientists tend to seek (e.g., Underwood, 1993).

Pagination is generally acknowledged to save time in the overall page-production process since it is faster to typeset the entire page as one element than to set each story, headline and caption on a page individually and then paste them up. If however, pagination requires more time in the newsroom and more editors are not hired, then, as Russial (1994c, p. 93) hypothesizes, depending on the amount of time that editors spend doing electronically what was once done manually in the backshop, the quality of the paper likely will suffer.

Newsroom management wants pagination because it gives editors greater control over the appearance of the page, and because it gives them greater control over meeting deadlines. Increasingly, the digitalization of all information—graphic and text-based—that appears in the paper or is collected by the newsroom is appealing because of the possibilities for adding value through "recycling" the information through other media such as fax and electronic newspapers, and audiotext and videotex products. Pagination has improved quality control and led to *better-looking newspapers*. The tradeoff, however, is *diminished content quality*, as well as editors who are spending less time on traditional editing tasks as they spend more time on production demands—with "a majority of those surveyed spending 20% or more of their time on former backshop functions" (Underwood, Giffard, & Stamm, 1994, p. 120). As with electronic editing, the more time editors spend on pagination, the less time they have for traditional journalism tasks.

Russial found that the "makeup" portion of the job—that which had transferred from the backshop to the newsroom—takes about twice as long as the more traditional editor's work, which is the creative page design process. Pagination systems are, at many newspapers, contributing to what Underwood et al. (p. 116) called a hybrid newspaper employee—one whose work combines the functions of copy editor, newspaper designer, and paste-up specialist.[4] One editor estimated that 60% of the work done on his design desk had once been done in the composing room, and in that instance, because photos were paginated as well as text, the pagination process took triple the time spent on page design (Russial, 1994c, p. 97).

Russial suggests that paginating editors felt that their workload was increased and that it reduced the time they had for "traditional editing tasks." Pagination also appeared to increase the proportion of essential tasks, and, in particular, those that were relatively mechanical or routine (Russial, 1994c, p. 98). Editors on deadline may be forced to reassess their editing priorities, asking themselves whether there is time, for example, to check facts, rewrite twisted prose, write a better headline, or monitor legal concerns. An editor's decision about this is an important element in the quality of the newspaper, since the copy desk is the last stop in the newsroom before the words on the screen become the printed facts which we comfortably construe as news and reality (Russial, 1994c, p. 98). The guidelines for making such decisions may be changing, since, as we've seen, editors are assuming more responsibility for the production work that determines at the most basic level whether the paper gets published. In light of the changing composition of

---

4. While it does not necessarily follow that this new organization of work will lead to fewer employees doing the same work, newsroom managers nonetheless acknowledge that these new jobs are designed to allow for that possibility.

their work, the more traditional emphasis of editors' tasks on accuracy and clarity may take a backseat to their production responsibilities.[5]

Management also expressed concern in some instances over the development of a "production mindset," noting that paginating editors seemed less eager to redraw pages between editions. "[Pagination] shatters a lot of the feel for journalism that editors [should] develop early on. I've seen it in action. ... People behave like a production department... I'm not sure what the solution is" (Russial, 1994b, p. 16). Russial also found that moving production functions into the newsroom replicated, between paginating and nonpaginating editors, tensions traditionally found between news and production. Pagination skills are increasingly a baseline requirement for hiring new editors, but editorial management may still be uncomfortable with the idea of a production functionality residing in the newsroom (Russial, 1994b). Some editors see pagination as "part of a relentless march in which technological imperatives continue to drive newspapers and override the journalistic mission of the newsroom," and to express concern that news organizations have become preoccupied with "the looks and style of the newspaper at the expense of content and journalistic substance..." (Underwood et al., 1994, p.124). Again, editors with less experience seem less troubled by the changes.

Computer-assisted reporting

While their traditional role may not have changed as much as editors', reporters nonetheless experience pressure to learn the whys and wherefores of new information technology. Of course, introducing VDTs to reporters also meant reducing the need for rekeyboarding stories, as well as lost jobs or retraining for the people in the backshop who had traditionally rekeyed them. And when there were computer problems, reporters found themselves frequently reliant on computer technicians who often had neither an interest in nor a clue about deadlines or journalism and its nobler pursuits in the interests of humankind. By the early 1980s, reporters were being drawn into a new level of computer use as newspapers increasingly turned the burgeoning stacks of clips, which tend to decay or become lost in various newsroom shuffles, into electronic libraries, or morgues.

---

5. Russial's (1994c, p. 98) study is useful in looking at the impact of pagination technologies on the amount of time and the type of work that editors are doing. However, as newspapers increasingly see themselves as "information companies" selling a broad range of products based on the same information database, pagination becomes not an end point in the information production process, but a stop along the way: Information, once in the digital form used for pagination, can be easily manipulated and repackaged for a variety of other media uses—including on-line access to newspaper libraries for both inhouse and public use, electronic newspapers, audiotext, fax newspapers, and Web pages. Thus, while it may take an editor as long to paginate a story as it did to have it composed by hand in the backshop, the newspaper benefits because the information remains in digital form and does not necessarily have to be reworked for other uses. Nonetheless, the more important question he raises addresses how a shift in their work composition affects the quality of the content.

As off-the-shelf personal computers (PC) clones began to replace proprietary VDTs in newsrooms,[6] breakthroughs in networking technologies and computer architecture made it possible to connect PCs to mainframes, and in many instances to replace them. In a 1984 survey, more than half of the 600 newspapers responding were making use of PCs somewhere in their operation.[7] Along with PC capabilities came the capability to put modems in the newsroom and allow reporters to dial up the burgeoning number of commercial and government databases available on line, and eventually through Internet connections. Koch (1991) argues that the primary tension involved in introducing electronic databases into newsrooms is between the apparent costs of the database systems' introduction and its benefit to the news gathering effort. But he also predicts changes in the content and narrative form of the news, as well as new forms of relations between "public information writers and their subjects." Journalists will gain greater flexibility, no longer bounded by "beats," leading, if you will, to more interdisciplinary coverage. This new organization of reporting will also change relations between newspeople, "who, in the past were divided by their respective responsibilities for a specific, topical specialty rather than united in their ability to focus on a story or issue" (p. xxv). He questions whether these "electronic resources" will "enhance or impede what might be called the 'minimal completeness and objectivity' criterion by which newspeople judge their own performance," but concludes that the derivative new form will finally allow "daily newswriters" to approach the standards they have "long asserted" (p. xxiii). He also predicts that newsroom managers should see improved productivity "as reporters are empowered to cover with increased criticality progressively broader classes of events" (p. xxiv).

Reporter use of electronic databases affords several potential benefits. It can make reporters less reliant on traditional sources; widen the considerations of their research; make more information available to them, more conveniently, in less time. It provides an opportunity to enrich reporters', and by extension, readers' understandings of issues. On the other hand, it is also expensive, and may draw resources from other, equally or more important, needs in the newsroom. As reporters spend more time with on-line research, they are likely to spend less time with other types, such as interviews and observations. Research shows that reporters have tendencies to become dependent on database sources, just as they have with official and expert sources. However, research does not extend to consider-

---

6. Newspapers had long felt that they were held hostage by traditional newspaper technology vendors who manufactured "closed" systems in which the entire system — mainframe, software, terminals, printers, and soon had to be bought from the same vendor. Systems from different vendors, say in classified and editorial, were incompatible. With PC-type and networking technologies, newspapers began to push vendors to create "open" systems, where relatively inexpensive PC clones could be used in place of "proprietary" VDTs, which could cost as much as $15,000 or more, each.

7. "Personal Computers are Booming but Problems are Looming," *Presstime*, July 1984, p. 45. Cited in Sohn, Ogan, and Polich (1986 p. 168).

ations of what social, economic or internal political pressures may contribute to this dependence. For example, do chain newspapers encourage or discourage reporters to rely on databases more often than newspapers with different ownership structures? And while interdisciplinary news may present greater flexibility, it may also represent the loss of specialized knowledge which arises from intimate and regular contact with a subject. The best possibility is that these and other computer-assisted reporting techniques are used to enhance rather than replace traditional news-gathering techniques.

There are additional possibilities, of course. For example, print journalists already conduct an increasing number of interviews by phone. Greater reliance on database reporting is likely to further reduce the face-to-face contact they have with sources. Reporters acknowledge covering a growing number of major events without going to the scene.[8] In increasing newsroom use of computer-assisted reporting, managers will need to look closely at the tacit knowledge that successful reporters bring to their work. Interviewing may be a form of professional knowledge that is essential, a skill that will be destroyed or lost to the extent both that it is less practiced, and that newsrooms rationalize the work of reporters.[9] Print reporters already compete with increasing numbers of media outlets, and are having increasing difficulty gaining access to public figures.[10] Speculatively, if interviewing skills declined, access would become even more difficult. Reporters may find that although they approach interviews with greater in-depth knowledge and better questions, they have less of the tacit knowledge that makes them able to persuade sources to talk openly with them. Journalist William Greider claims that, already, "Most reporters aren't interested in finding out what happened. Most journalists are interested in finding out about news, which is another commodity, one interested in what's the angle today, or this week or tomorrow.... They want to plug a hole in their story or give their pieces a voice to make their point" (Rosenstiel, 1995, p. 24).

Because of the costs of and the need for technical skills demanded by electronic research, reporters may have less control over what information is available to them than before. Research shows that they typically perform narrow searches that offer them information rather than understanding. The training they receive, in most instances, is designed to teach them basic commands, not the more complex skills needed to use the new technology as a means of developing a broader and more deeply contextualized news account. In other words, this kind of practice is likely to exacerbate the trend of treating events as discrete rather than linked, despite the potential that would allow opportunities for what many critics would see as "better" journalism, although Koch shows more optimism in his predictions for

---

8. Robert Scheer quoted in Rosenstiel, pp. 23–24.

9. Zuboff, p. 186, uses Michael Polanyi's description of tacit knowledge as "forms of meaning [that] are comprehensible only as a whole and can be destroyed when objectified and analyzed."

10. Scheer quoted in Rosenstiel, p. 26.

interdisciplinary beats and "empowered" reporters capable of greater productivity. The set of characteristics that can make a news product "better" is enormous. To date, conclusions about better newspapers through better technology have often accepted unchallenged the link between technical improvements in speed, accuracy, productivity and better news products. As applied to the latter, that link is highly suspect. As Weaver and Wilhoit (1986) point out, "Even though the bulk of U.S. journalists perceive new technological devices as improving the quality of their work and saving time, it may be that in the long run such devices will change the nature of the messages created by these journalists in ways that do not benefit the society at large" (p. 159).

Depending on the balance between on-line database and royalty costs versus labor costs, managers also may find that, just as it is often less expensive to publish syndicated columns and features than to pay staff to write (Conniff, 1994),[11] it may be less expensive to draw research from on-line sources than it is to send reporters out to cover all but the most high-profile local stories. Although editors argued that "good reporting habits as traditionally taught" would overcome the downside of database use, it is clear from the experience of both copy and pagination editors that journalists do not always have discretion over their use of traditional skills. Journalism schools are also under increasing pressure to ensure that graduates have adequate technical skills by the time they enter the newsroom (e.g., McClain, 1994; Russial, 1994a). Unless students are required to take more courses, students may come to newsrooms with less knowledge of traditional skills and practices. As in other jobs in the newsroom, there may be a tension between the demand for technical versus communicative skills.

It seems unlikely that databases will provide greater objectivity, since although information will be stored, retrieved and reproduced in a different form than before, it will continue to be gathered by humans. But increasing reliance on a narrower range of sources could lead to declining quality and standards of reporting and an increasing appearance of news stories which are little more than continually redigested and reinterpreted pieces of assembled data.

NEW FORMS OF WORK ORGANIZATION IN THE ELECTRONIC NEWSROOM: THE END OF JOURNALIST SPECIALIZATION?

Newsrooms—and society generally—have gained more experience with technology, as well as a greater sense of its unexpected, or "unintended," effects. A new generation of newsroom managers includes former reporters and editors who underwent many of those changes. At the same time, newspapers are calling them-

---

11. For example, Conniff (1994) writes in a column on newspaper-telephone company alliances, "Much of what many newspapers provide today is canned corn to begin with: a frequently bland diet of syndicated news, features, cartoons and columnists."

selves information companies or information providers, and newsrooms are colliding with marketing departments and learning to think about how they can "add value" to the traditional news product. In this process of self-reconceptualization, newspaper companies have begun to reorganize the newsroom into teams, a form of work organization first linked to the flexible manufacturing processes in computerized industrial plants.

Newsroom work teams take several forms, often called design desks: Some focus primarily on pagination and design, while others stress the creation of a "better packaged story." Whether in newsrooms or manufacturing facilities, cross-training, so that employees are "interchangeable," is implemented either as a part of, or in conjunction with, the team structure. Russial's (1994c) study of newsroom pagination and work organization (pp. 20–21) found that newspapers are still experimenting with new forms of staff organization and job design. Technological deployment may be creating new organizational problems, for example, lack of leadership, skilled staff or other resources, and possibly limiting a newspaper's ability to change in other ways.

Many newsroom design desks handle layout and pagination almost exclusively, and are implemented to "control the look of the paper, speed and cost" (Russial, 1994c, p. 13). Design desks often lead to better looking pages and greater uniformity throughout the paper, as well as improved coordination among editors, designers, and reporters. Editors sometimes report feeling more creative. Often, papers with design desks have created new job classifications to handle the reorganization of work. Page designers, who have more newspaper experience and may be paid comparably to other editors, laid out pages and sections, while design assistants, who function as technicians and for whom newspaper experience may be considered unnecessary, may be paid substantially lower wages. One news executive was quoted as saying, "We don't want to pay someone $800 a week to 'design' those holes. We want to get someone who can mass produce" (Russial, 1994c, p. 14).

Some other papers organize work so that more technically adept editors handle pagination, leaving other, perhaps less adept editors to more traditional tasks. Russial found that many editors acknowledged that it may be important for pagination to be done by editors, despite its technical nature, since "most editors felt more comfortable with other editors" making decisions about necessary adjustments such as trimming a story or photo or adjusting the layout. One newspaper based its decision about which pagination system to buy on the belief of editorial management that "it would allow the production function to be shifted to news professionals with the least degradation of their job and with the greatest opportunity to preserve editorial control over the many little decisions that must be made to implement the larger decision of drawing the page" (Russial, 1994c, pp. 15–16). Russial's study indicates that newsrooms feel some tension between, on the one hand, rationalizing design desk activities so that editors handle tasks which require high-level skills while technicians are assigned more routinized work, and on the other, recognizing the value of professional judgment and retaining maximum levels of editorial control.

Work team strategies also allow management to construct a newsroom more responsive to marketing needs by moving the newsroom away from its traditional assembly line structure in which "stories are passed from reporter to editor to production with little communication in between. This structure facilitates the need to meet daily deadlines, not necessarily what works best for readers" (Auman, 1994, p. 129). For example, the "maestro approach" is, from its inception, a means of understanding what stories are important to readers and then matching the skills of the newsroom to the challenge at hand (Ryan, 1993, p. 21). It is intended to promote stories that evolve out of a process that focuses on readers' questions (p. 22). The maestro, or team leader, directs, orchestrates, and referees, as well as reserves space and color, or "whatever it takes to tell the story in the most meaningful way for readers" (p. 21). Auman also found that newspapers implement team strategies to create a management tool that make newsroom employees and machines more effective in the overall activity of telling stories by creating teams of copy editors, photographers, page designers, and reporters. Underlying this need is increasingly stagnant circulation which drives newspapers to seek new ways to attract readers, with many deciding that better organization and presentation of information are key.

Under ideal circumstances, teams can, as Auman suggests, restructure the newsroom production process in a way that brings together the issues of organization and presentation—integrating writing, editing and design; but she also questions whether, in the process, content is subverted to art and presentation. At their best, teams reorganize traditional tasks to take advantage of new information technology potential as well as human creative potential. Yet depending on their makeup, they may emphasize design over content and cede control of the news to the design desk, and thus may represent a threat to traditional standards of quality based more on content than appearance.

In some newsrooms, design desk employees are cross-trained on a variety of tasks, so that "word people acquire visual skills, and visual people acquire word skills" (Auman, 1994, p. 139). To fill these positions, editors seek journalists with layout and design skills first, and news judgment second. Large papers rank Macintosh computer graphics skills third, but papers with circulation under 100,000 placed a priority on more traditional journalistic skills, although many of those papers plan to use Mac-based pagination systems. The successful design desk editor, concludes Auman, "is perhaps a new breed of journalist who has an integrated editing mind—who is a word person but who can integrate words with visuals" (p. 140). It is also clear, however, that technical skill continues to rank high.

Auman's study found that editors on design desks spent about half their time on page design and layout, 15% on pagination, and the remainder on tasks such as headline or cutline writing, creating infographics, photo editing, and coordinating people and elements in a story or project on a page—with some variation in task depending on the size of the paper (p. 131). Among these tasks, the more time a design desk spent on creating graphics, page design, and pagination, and the less spent on headline writing, the more likely editors were to rate the desk a success. This is reflective of the hiring priorities that newspapers have established: Editors whose primary skills are in design and layout seem more likely to focus on presentation over content. This type of value system would have been nearly inconceivable prior to newsroom implementation of desktop publishing technologies which provide flexibility in manipulating page elements and executing more complicated design concepts.

Applebaum and Batt (1994) found that in industrial worksites, systems of work organization similar to the above often include only a few select workers; the number of participants in workplace teams who have actual authority to make decisions is often limited to an upper echelon, already more empowered than most workers. The structure of the newsroom teams replicates this by choosing primarily newsroom leaders with seniority as the team leaders. While other workers may have input, it seems likely that they ultimately have less influence than the leader.

One aspect of the logic behind design desks, however, seems to be to continue a trend toward automated production work that has been moved to the newsroom, which has already created a production mentality among some editors. By separating out the less skill-intensive work, managers are creating new entry-level work in the newsroom, sometimes and sometimes not an opportunity to develop skills or to find a stepping stone into better paid and more skill-intensive positions. Managers recognize the importance of professional judgment and have designed the desks, in many instances, in ways that leave the traditional hierarchical aspects of organizational decision-making intact. Newsroom teams, in some incarnations, may provide an environment that allows workers to broaden their skills and to contribute more creatively to projects, but these types of newsroom variations on teamwork remain nascent.

CONCLUSION

News is a social construction rooted in journalistic routines and practices, and newspapers are central tools we use to construct our own individual and collective senses of reality. Decisions about how to design and deploy information technology have changed some routines and practices of journalists at U.S. newspapers, and transformed understandings of what basic journalistic skills are. Journalists' work has traditionally required communicative and coordinative skills and the manipulation of symbols. Social choice about technological deployment and design have led some workplaces to replicate the 19th-century logic of automation, in which work is alienating, meaningless to workers, and gutted of opportunities to show any but operational skill. Other choices have led to workplaces that encourage workers to develop new intellective skills and creativity in dealing with technological change. Technological change occurs in both social and historical contexts which exist both in and outside the newsroom. Issues as diverse as concern over social status, and interests, beliefs, and values about what constitutes meaningful work and even the extent to which work should be meaningful, as well as what problems technology can or should solve, all influence whether a workplace takes one direction or the other. Often the people who make decisions about the purchase and deployment of technology come from very different backgrounds ranging from data processing to composing to journalism, which all offer very different perspectives on how and toward what ends work should be organized.

During the last 20 years, computers and computer skills have become integral to getting the news out, and editors, relatedly, have acquired increasing responsibility for production processes. In doing so, the emphasis on traditional journalism skills has shifted to share predominance with computer skills, which often reorients the nature of work from sentient toward abstract, sometimes shifting emphasis from subtle features of journalistic accuracy to the more immediate allure of technical precision. Editors' work is often more stressful than it was previously. They frequently need more skills, have a broader range of duties, less backup support, and earlier deadlines for more stories and more products—whether those products are zoned editions or Internet editions. The shift from sentient to abstract work, increases in the pace and quantity of information flow into and within the newsroom, and the additional responsibility of producing divergent products also have increased the pressure of this work in many contexts. Often, additions to staff have not been commensurate to increases in breadth and quantity of work.

Many workers, unable to affect the process of change, have responded by leaving the newspaper or the industry, taking with them their skill and experience, or by becoming demoralized and halfhearted in their work. They felt that the procedures of "technojournalism" were robbing them of the ability to produce a newspaper that met their expectations of quality, and leaving them with a technical skill that had little coincidence with their primary interest in the profession. Recent downsizing trends in the industry are likely to intensify and widen this response. Clearly, in such instances, it isn't only journalists who lose: Newspapers lose the contributions of motivated and engaged journalists, and they lose the synergy that comes from creative and sometimes casual human interactions. However, research into the effects of both team production and older methods of work organization indicate that unions have been an important factor in retaining quality production standards and protecting worker rights (e.g., Applebaum & Batt, 1994; Shaiken, 1986; and to a lesser extent Zuboff, 1988). Weaker union structures present less threat to management control of the labor process in the newsroom than the strong printers unions whose power was diminished with changes in production technology. As research into these questions continues, it will be important to evaluate the extent to which the same holds true in newsrooms.

Research into the deployment of information technology into U.S. newsrooms in forms of electronic editing and pagination shows that goals have been to reduce staff and increase profit and efficiency—to replace human bodies with machines, to speed processes, and to gain greater control over the labor process. The unintended consequences have reduced opportunities to utilize existing skills, and left work meaningless and workers without control over their work; worker demoralization and loss of product quality often accompany these consequences. This is not inherent in the technology, but is a result of the social choices that surround its design and deployment.

As computerization of newsrooms proceeds, it will be important to investigate whether new forms of work organization in the newsroom will provide journalists with the opportunities necessary to develop critical judgment and analysis skills that have traditionally been a primary requirement of journalistic practices. To date, available research shows that editors report a loss of traditional skill, and a more depersonalized and abstract environment. The direction of computer-assisted reporting remains ambiguous. Journalism schools are under increasing pressure to ensure that graduates have adequate technical skills by the time they enter the newsroom. Unless students are required to take more courses, students may come to newsrooms with less knowledge of traditional skills and practices.

Will these newsrooms attract or keep people to whom the use and development of high-level versus operational skills is important? Are high-level skills fundamentally at odds with the prevailing logic governing the deployment of information technology in newsrooms? Among the consequences of the logic of automation is that employees are eliminated rather than challenged to participate as full members of the organization with equal stakes in improving quality and imagining new production possibilities. Despite the creation of derivative products and the need for new employees to produce them, the number of people required to produce *any given item* continues to decline with automation (Zimbalist, 1979, p. 110). Often, experienced workers are seen as liabilities rather than assets. Those who remain in the newsroom may receive the training, education, opportunities, and structure that will help them develop the intellective skills that will enrich their jobs and add value, but the specter of unemployment will remain a destabilizing and corrosive undercurrent in the newsroom and in the news industry.

This has implications for society, of course, regardless of the product. But it matters more if the product is a newspaper, if the number of voices who participate in constructing this particular "symbolic consumer product" declines, whether through departure from industry or because the work itself begins to quiet and dull journalists' curiosity and passion. While new opportunities for participation arise out of new technological and media forms such as the Internet, even here it is likely that the logic of capital and accumulation will influence, if not shape, social choices in a manner that constrain the liberating potential of these new forms. The set of characteristics that can make a news product "better" is enormous. To date, conclusions about better newspapers through better technology have often accepted unchallenged the link between technical improvements in speed, precision, productivity and better news products.

Ultimately perhaps, the most central issues that these explorations will help to illuminate is what impact the reshaping of newswork and the reorganizing of American newsrooms will have on the final product—the information by which a community knows itself, and on which it bases its judgments. Since news is a social construction (Tuchman, 1978) that reflects not only society, but newsroom practices, it seems unlikely that a largely automated newsroom will be able to respond with spirit to the challenge thrown out to it in 1861—"It is a newspaper's duty to print the news and raise hell."[12]

REFERENCES

Applebaum, E., & Batt, R. (1994). *The new American workplace: Transforming work systems in the United States*. Ithaca, NY: ILR Press.

Auman, A. (1994, Spring). Design desks: Why are more and more newspapers adopting them? *Newspaper Research Journal, 15*(2), 128–142.

Braverman, H. (1974). *Labor and monopoly capital: The degradation of work in the twentieth century*. New York: Monthly Review Press.

Conniff, M. (1994, December 17). The leading edge: The enemy is us. *Editor & Publisher, 127*, 21, 43.

Deetz, S., & Mumby, D. K. (1990). Power, discourse and the workplace: Reclaiming the critical condition, *Communication Yearbook 13*, 18–47.

Diamond, E. (1993). *Behind the Times: Inside the New York Times*. New York: Villard Books.

Gerbner, G. (1995, January/February). Response to a technology column by Stephen D. Isaacs, "The age of iron pyrite, maybe?" *Columbia Journalism Review, 33*(5), 5.

Koch, T. (1991). *Journalism for the 21st century: Online information, electronic databases, and the news*. New York: Praeger Publishers.

Kurtz, L. D. (1980, Summer). The Electronic Editor. *Journal of Communication, 30*(3), 54–57.

Lindley, W. R. (1988, Summer). From hot type to video screens: Editors evaluate new technology. *Journalism Quarterly, 65*(2), 485–489.

McClain, R. (1994, January). Journalism education should include more computer training. *Newspapers & Technology, 5*(12), 21.

---

12. Quote is from the *Chicago Times*.

Rosenstiel, T. (1995, January/February). Yakety yak: The lost art of interviewing. *Columbia Journalism Review, 33*(5), 23–27.

Russial, J. T. (1994a, August 13). *Beyond the basics: Mixed messages about pagination and other skills.* Paper presented at the Joint Session of the Communication Technology and Policy and Newspaper Divisions of the AEJMC, Atlanta, GA.

Russial, J. T. (1994b, August 13). *Pagination and newsroom organization.* Paper presented at the Joint Session of the Newspaper Division and the Communications, Technology and Policy Division of the AEJMC, Atlanta, GA.

Russial, J. T. (1994c, Winter). Pagination and the newsroom: A question of time. *Newspaper Research Journal, 15*(1), 91–101.

Ryan, B. (1993, March). Editing takes on a new look. *Quill, 81*(2), 18–24.

Shaiken, H. (1986). *Work transformed: Automation and labor in the computer age.* New York: Lexington Books.

Sohn, A., Ogan, C., & Polich, J. (1986). *Newspaper leadership.* Englewood Cliffs, NJ: Prentice-Hall.

Thomas, R. J. (1994). *What machines can't do: Politics and technology in the industrial enterprise.* Berkeley, CA: University of California Press.

Tuchman, G. (1978). *Making news: A study in the construction of reality.* New York: The Free Press.

Underwood, D. (1993). *When MBAs rule the newsroom.* New York: Columbia University Press.

Underwood, D., Giffard, A. C., & Stamm, K. (1994, Spring). Computers and editing: Pagination's impact on the newsroom. *Newspaper Research Journal, 15*(2), 116–127.

Weaver, D. H., & Wilhoit, G. C. (1986). *The American journalist: A portrait of U.S. news people and their work.* (2nd ed.). Bloomington, IN: Indiana University Press.

Zimbalist, A. (1979). Technology and the labor process in the printing industry. In A. Zimbalist (Ed.), *Case studies on the labor process.* New York: Monthly Review Press.

Zuboff, S. (1988). *In the age of the smart machine: The future of work and power.* New York: Basic Books Inc.

10

Journalists' Use of
On-Line Technology and Sources

Steven S. Ross
Columbia University

At all stages of the news process—story conception, research, finding and interviewing sources, establishing source credibility, information analysis and presentation, even distribution of finished articles—journalists are using computers in increasing numbers. This is despite the many technical barriers that have made online access difficult in news organizations. Large newspapers, especially, place typesetting terminals on journalists' desks. These terminals are critical production tools: direct connections between the internal production-oriented network and outside networks are technically difficult and potentially dangerous. These connections have only recently begun to appear, allowing journalists to move seamlessly between the Internet and other on-line networks, and internal production systems.

It is clear, however, that the major barriers are technical. There are, for instance, no obvious differences between male and female reporters' use of on-line services when they are on the same beat. Perhaps the stereotype is because such services have tended to be more heavily used among sports and business reporters than among journalists on other beats. And sports and business sections tend to have higher percentages of men on reporting and editing staffs.

There are also no obvious gender (or race) differences among those using analytical software such as spreadsheets and relational database programs for computer-assisted reporting. Women and minority-group journalists author many such stories. At Columbia, where we have been teaching mandatory spreadsheet and database use to all students since 1987, we see absolutely no difference in student course performance by gender or race.

But all that is computerized is not analytical, or even reporting. Journalism jobs that are specifically in new media organizations are often glorified production positions, especially when those new media organizations are associated with newspaper Web sites. The field is still evolving.

This chapter concentrates on print journalism, especially in newspapers. It includes data on magazines (and some asides about broadcast) for comparison.

With Don Middleberg, head of Middleberg & Associates, a New York-based public relations firm, I conducted two large surveys of print reporters—one in August 1994 and one in September 1995—to quantify these gains. These surveys, of 6,000 and 3,800 respectively, are the largest of their kind. They generated more than 1,500 usable responses—high for mailed surveys. Various tests, described later in this chapter, allow us to assume that the responses mirror the overall sample. All U.S. daily newspapers were queried in each year. The 1994 survey also included all U.S. weeklies with circulation larger than 2,500. There were 1,500 magazines surveyed in 1994, and 2,000 in 1995. The magazines were randomly selected from the nation's 14,000 generally circulated titles (avoiding house organs). Sampling error is detailed later in the chapter. But, in general, reported percentages for the overall sample each year are plus or minus 4% at the 95% confidence level.

I have also conducted numerous samplings of broadcast and print reporters in other settings with nonrepresentative samples, to help frame questions for the main surveys. Insights gained from these efforts, from on-site examinations of media operations, and from watching the progress of our graduates, form the basis for this chapter.

To summarize the "good" news, assuming that greater access to a broader variety of information sources inherently makes for better stories: There has been an enormous increase in print journalists' use of on-line sources of late. Almost a fourth of the respondents (23%) to the fall 1995 survey say they or their staffs use on-line services daily. More than two thirds say they use such services at least once a month. Those totals were significantly higher than for the survey taken in the summer of 1994. The growth has been mainly in daily newspapers; monthly (or more frequent) use by newspaper staffs rose from 44% to a startling 71% in the 14 months between the two surveys.[1] In this chapter, where inclusion of weeklies in the 1994 survey distorts year-to-year comparisons, we compare only dailies to dailies each year.

Despite press reports predicting the demise of value-added on-line services in the face of Internet growth (and especially growth of the Internet's World Wide Web), the press itself favors value-added commercial services for its searches. In 1994, the leader in our survey was CompuServe. In the survey conducted in the fall of 1995, to the surprise of no one who has been following America Online's

---

1. Both values are +/- 4%, so the difference could be as large as that between 40% and 75% and as small as 48% to 67% at the 95% confidence level. Various limitations in verifying randomness would tend to lower the possible interval further; various internal self-consistency checks in the questionnaires themselves tend to more than make up for any lack of randomness.

(AOL) marketing blitz, AOL surged ahead. By fall 1995, twice as many journalists responding to our survey ranked AOL first (40%) for commercial services and Internet access, compared to the Internet alone (20%). CompuServe ranked in third place by fall 1995, at 19% (it had been first, at 35%, in 1994). Of course, many use AOL to access the Internet despite its high cost relative to local providers, because it is easy to set up and obviously available. This tends to inflate AOL's scores.

The situation is likely to remain fluid for some time. Since the fall 1995 survey was completed, AOL and CompuServe have stepped up their offers of free accounts to news organizations. Both have also improved access to the Internet from inside the commercial services. AOL even is expected to offer limited access to AOL features from the World Wide Web (WWW).

Once editors and reporters go on line, they are interested in gathering reference materials, finding raw data (in particular, government figures and legislative doings, and corporate biographies), using e-mail and seeking out new sources for interviews. In that last function, editors and reporters will often use old-fashioned on-line Internet utilities such as Listservs and Usenet newsgroups, as well as the WWW.

One might think that younger, more computer-literate journalists would be leading the charge into the computer age. There is some anecdotal evidence, and evidence from our nonrepresentative sampling at professional meetings, that this is so with respect to use of on-line services. There is no evidence, however, that this is so with respect to use of analytical techniques such as spreadsheets. A quick look at bylines suggests, in fact, that older journalists use numeric analysis more than younger ones do. This may be due to the increased responsibility which older, more experienced journalists have for major analytical stories at news organizations—or due to poor math and logic skills among recent graduates!

The 1994 survey was the largest ever for computerization of print media—more than 6,000 writers and editors were queried. The 1995 survey concentrated and expanded upon coverage of magazines and of daily newspapers. Thanks to the exceptional response rate (about 500 of 2,000 magazines and more than 300 of 1,815 newspapers in 1995—almost all of them dailies), we were able not only to track overall use of on-line services, but also to gain a better understanding of how journalists use these services. The magazine section of the 1995 survey was the largest ever conducted of magazines in cyberspace. Responses to the newspaper section of the 1995 survey cover 40% of U.S. daily newspaper circulation.

The 1995 survey calculations are based on 751 responses, out of more than 800 received (some responses lacked answers to questions on circulation and frequency, and thus could not accurately be matched to the 1994 cohort). The 1994 survey results were based on 725 usable responses, also out of about 800 received.

Although we have informally sampled broadcast media, and found substantial growth in on-line services provided by and used by broadcast media, we see no consistent pattern in such use. That is, outside the networks, whose on-line use is growing rapidly, we see no particular pattern between a broadcast or cable outlet's adoption of on-line services for research or for disseminating add-on services, and the outlet's staff size, market size (as measured by Arbitron's Area of Dominant Influence and share), or years in market. Instead, entirely new services are springing up within broadcast and cable markets. And in traditional organizations, the existence of a "champion" seems more important than other variables. By "champion," we mean an individual who understands the process and economic benefits of going on line and who is willing to press for including an on-line component in the organization's media product lineup.

ON-LINE REPORTING: THE DETAILS

As already noted, 23% of those surveyed in fall 1995 said they or their associates use on-line services roughly once a day or more. Another 24% said they use such services at least once a week on average. Overall, 68% said they or their staffs use on-line services at least once a month.

This reported level of use was up considerably from the 1994 sampling, which showed that only 50% of print media journalists used on-line services at least monthly. Both values are plus or minus 4% with a confidence of over 95%. So at worst, the growth was from 54 to 64%, or roughly a 20% annual growth rate. More likely, the annual growth rate was closer to 35%.

Figure 10.1: How often do journalists search for information on line?

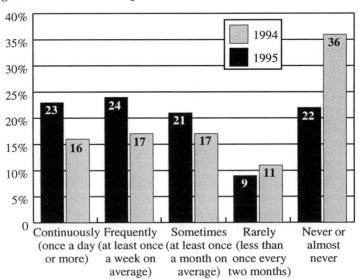

The growth was wholly attributable to newspapers, not magazines. For magazines, monthly use of on-line services was steady within the statistical limits of the survey, (69% in 1994, 65% in 1995). For daily newspapers, however, it rose dramatically, as already noted, from 44% to 71%. (See methodology, below, for discussion of sample differences.)

What's more, respondents from newspapers (and their staffs) were much more likely to use on-line services at least weekly by fall 1995 (56%) than summer of 1994 (22% for the entire newspaper sample, 31% for dailies).

Although the Internet gets most of the publicity, and although pundits predict the eventual decline and or demise of commercial services such as CompuServe and AOL, journalists overwhelmingly use these value-added commercial services. Internet use is obviously expanding, however. By the time this book appears, CompuServe and AOL should both be offering more seamless connections to the Internet, especially to the WWW. In fact, the relative position of all commercial services except AOL dropped from 1994 to 1995. Direct Internet use by journalists increased somewhat. This use has probably been accelerating since the 1995 survey; AT&T and MCI are among many communications giants that started low-cost access services to the Internet in the past year.

Most journalists who access the Internet do it through a commercial value-added provider such as AOL, CompuServe, or Prodigy—and not through a local or national Internet "provider" that offers an Internet on-ramp but little or no value-

Figure 10.2: Which on-line services do reporters use?

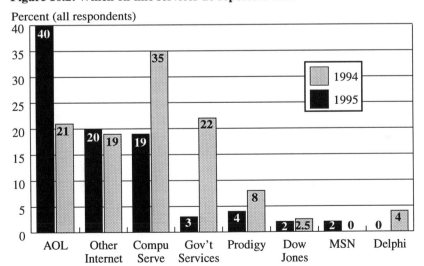

added services. Local "on-ramp" providers are much more likely to be used than are the few large national or regional providers. During the fall 1995 survey period, there were about 600 local providers in the United States. Newspapers were beginning to become local Internet providers themselves. They ranged from the *Waterbury (Connecticut) Republican-American* to *The Washington Post.*

One third of the Internet users in our sample accessed the Net through their employer's host. News organizations are probably similar to other organizations in this regard. That is, other studies have shown that employees of organizations that have their own hosts tend to use the Internet more often because the extra cost of using it, once a host is in place, is minimal.

More than two thirds of the 1995 respondents said they used the Internet (69%, or 519 out of 751 respondents) for one reason or another. This is a larger number than those who said they *or their staffs* used the Internet frequently. (Remember, 68% said they or their staffs used on-line services, including the Internet, at least once a month in the 1995 survey, and 50% said they did in the 1994 survey.) The responses suggest that knowledge of the Internet is higher than "frequent" use by a fairly wide margin. What's more, almost all who said they use it, do so for both personal and business reasons. The bulk of Internet use by journalists is for article research, e-mail, and finding sources or experts, but this is not the same as saying that 69% of all print journalists use the Internet specifically for journalism.

One fourth of those who use the Internet said they do so to retrieve press releases, essentially replicating electronically a traditional journalistic task. More than a third of those who use the Internet use it to access Usenet newsgroups—something that is possible on the World Wide Web, but as easy or easier to do in a text-only system.

In both surveys, editors judged on-line material provided by nonprofit or governmental organizations to be generally more reliable than material provided by business. While the gap is small, it's significant. Written comments volunteered by respondents suggest that one reason involves perceived comprehensiveness of government-supplied data. Reports by nonprofit groups—even activists—are often based on government-supplied data as well. In general, nonjournalist Internet users expect comprehensiveness and inclusiveness (one measure of "fairness," if you will) in Internet postings.

Some comments by respondents to this study, and some more general studies of journalists' sourcing habits, suggest that journalists are more leery of "business," because profit-making organizations have a more vested interest than do nonprofits and government agencies in getting their points of view across. Journalists should, of course, take nothing for granted in any data, no matter what the source. Nevertheless, the survey results specifically suggest that businesses placing material on-line for editorial use should be extra careful about accuracy and errors of omission; credibility is key—and fragile.

Figure 10.3: How do you access the Internet?

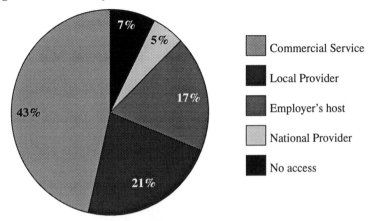

- Commercial Service
- Local Provider
- Employer's host
- National Provider
- No access

Figure 10.4: Where do you access the Internet?

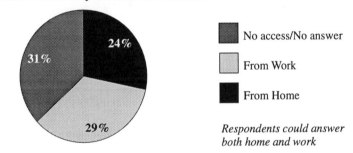

- No access/No answer
- From Work
- From Home

Respondents could answer both home and work

Figure 10.5: What do you use the Internet for?

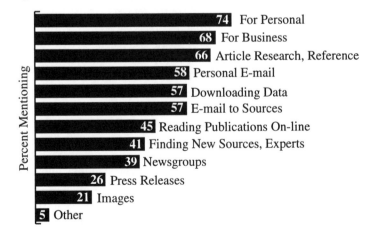

74	For Personal
68	For Business
66	Article Research, Reference
58	Personal E-mail
57	Downloading Data
57	E-mail to Sources
45	Reading Publications On-line
41	Finding New Sources, Experts
39	Newsgroups
26	Press Releases
21	Images
5	Other

Percent Mentioning

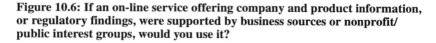

Figure 10.6: If an on-line service offering company and product information, or regulatory findings, were supported by business sources or nonprofit/public interest groups, would you use it?

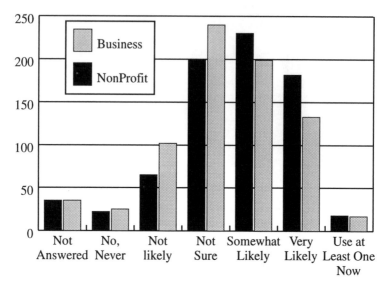

About 46% of the respondents said in 1995 they were at least "somewhat likely" to use business services on-line. That's down slightly from 49% in the 1994 survey, but the difference is not statistically significant. When asked if they would use services provided by nonprofit or public interest groups, 57% said in 1995 they were at least somewhat likely to do so. That's down slightly from 1994's 59%.

On the flip side, only 3% said in 1995 they would flatly refuse to use such a service. The few scattered comments from the 20 or so "refusers" in the 1995 survey suggest that they may like to report only from live interviews, and not from documentary sources of any kind. The chart below provides actual numbers of respondent answers for the 1995 survey. Of the 751 in the analyzed sample, 35 did not answer this question.

The key uses cited by our respondents in 1995 for such on-line services provided by business or nonprofit organizations would be for *reference* (mentioned by more than 40% of the respondents), press releases (mentioned by more than 34%), and raw data such as Securities Exchange Commission (SEC) filings and Environment Protection Agency (EPA) material (mentioned by more than 27%). In other words, raw data and source documents have the most pulling power. Images are gaining in importance.

The survey itself did not address the issue of how, exactly, journalists treat information from Internet sources that are less formal, such as e-mail, Usenet newsgroups, and Listservs. The problem with such "sources," is that reporters have no

Figure 10.7: Journalists' use of on-line services supported by business or nonprofits

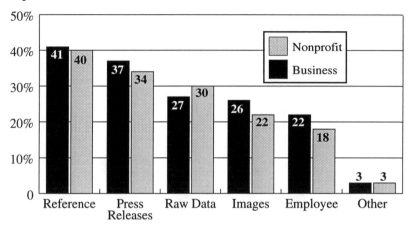

sure way of knowing the true identity of the source. That 60-year-old professor a reporter finds on a Listserv or in a newsgroup posting may really be a 13-year-old seventh grader. An e-mail address that looks genuine may have been forwarded through Fidonet or a third-country Internet server specifically set up for the purpose. Such forwarding is easy to do on the Internet or through a telephone-accessed electronic bulletin board. In the forwarding process, the e-mail address of the original sender is stripped away and replaced with a fake.

In responses gathered at professional meetings, reporters have been near unanimous in their insistence that firm identification be achieved before such sources are quoted. This, they say, would be achieved by, at the very least, an exchange of phone numbers followed by the reporter's call. Follow-up communication, once the source is interviewed by phone or in person at least once, could be by e-mail, respondents say. The chain of questions and responses is much harder to fake.

Nevertheless, many stories do find their way into print quoting unverified e-mail sources. It can be argued that many reporters do not understand the potential for fraud. But many of the stories involve reporters who **should** understand; the stories are written by computer columnists on computer-related topics, sometimes for computer magazines.

There are not many alternatives to on-line services and the Internet for distributing large amounts of current information. CD-ROM use, for instance, greatly expanded in 1995 (most new computers shipped for home use included CD-ROM drives)—but not by enough to make it worthwhile to send editors CDs unless you know they can read it at their desks. In 1994, only 17% of the respondents said CD-ROMs could be read at editors' or reporters' terminals. By fall 1995, that hadn't changed much overall. (The survey indicated 15%, but the results were

Figure 10.8: Where can CD-ROM disks be read?

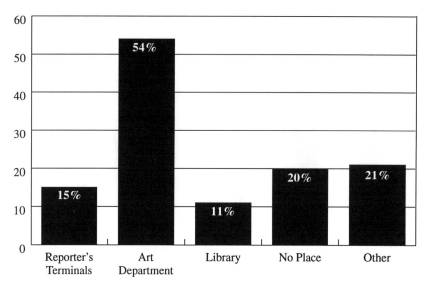

lower in 1995 mainly because freelancers were not included in the 1995 survey; freelancers working at home were far more likely than the norm to have CD-ROM drives.)

However, availability of CD-ROM drives in newspapers' art departments increased dramatically, by 22 percentage points from 1994 to 1995. In the 1995 survey, 54% said CD-ROMs could be read in the art department. In the 1994

Figure 10.9: Do journalists use press releases on disk or "infobases"?

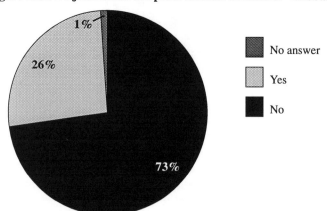

survey it was 32%. This is in line with increasing demand for art to be supplied electronically rather than in the form of a slide or reflective art.

Only 26% of the 1995 respondents said they use press releases on disk, or "infobases," however. These are databases contained on a floppy disk or CD-ROM. That's up slightly from 1994's 22", but not by a statistically significant amount. Among magazine respondents in 1995, 21" used infobases, compared to 18" in 1994.

Slow growth in this area has not matched expectations. Although providers of on-line information, such as businesses or government agencies, may want to provide "databases" that contain such materials as catalogs, organizational directories, and laws, the media have not been particularly interested in the extra detail that electronic delivery has made possible. Also, users complain that each infobase has a different interface that has to be learned.

Providers of infobase software, such as AskSam, Folio, ZyIndex, and MicroRim, say the trend may be toward allowing users to "search" an infobase through the WWW, rather than sending users a disk or a CD-ROM. Folio, MicroRim, and AskSam all have "Web server" products designed to do just that.

LIBRARIES AND LIBRARIANS

Media use of reference materials is generally weak in other ways as well. Most important, library services in media organizations are shamefully inadequate. On-line services can help reporters access such reference materials. But you can't send a disk to a media librarian and expect the information, or availability of it, to become common knowledge in the organization, because there may not be a central repository to receive and catalog the disk in the first place. The 1995 survey confirms that news libraries are an endangered species among magazines, and becoming more so. But newspapers began improving their library staffs in 1995 compared to 1994.

Only 27% of the respondents in 1995 said their organization has a news library with a designated librarian, compared to 23% in 1994. The total for those without a news library at all was unchanged at 47%. The remainder had news "libraries" that are essentially unstaffed, or supervised only by a secretary. A typical news library for a small publication would include back issues or clips, and standard reference works and directories.

The sample for the 1995 survey, however, omitted most weeklies. Among magazines in 1994, 17% had a library with a designated librarian, compared with 10% in 1995. Among dailies, the rise was from 27% to 61%. This suggests that as they receive more material on line than they have in the past, and as they seek to post material of their own, daily newspapers have started to re-invent and to re-invigorate their library systems. A quick nonrandom check of 12 medium-size dailies with news libraries and designated librarians suggests, however, that not all of the

Figure 10.10: Do media organizations have a news library?

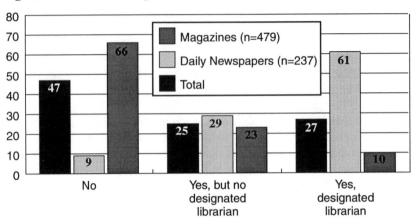

"designated librarians" are actually fully trained "traditional" library science/library service personnel; many may be trained only in database or Internet access. Only one of the 12 dailies (circulation 25,000 to 60,000) had actually employed a library-school graduate to oversee its library.

Figure 10.11: Do you expect your publication, or sections of it, to be distributed electronically?

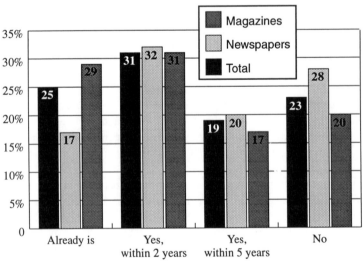

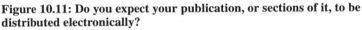

All respondents

PRINT MEDIA ON LINE

It should come as no surprise that newspaper and magazine organizations' plans for electronic distribution exploded in 1995. Some 46% of the 1994 sample had no plans to go on line a year ago. That percentage was cut precisely in half, to 23%, in the interval between the two surveys. Some 29% of magazines in the sample were distributing electronically by fall 1995, and another 31% expect to be doing it by fall of 1997.

Among dailies, all but 30 % expect to be on line within 5 years; 15% already were by fall 1995, and half plan to offer on-line products by fall of 1997.

Among daily newspapers in the fall of 1995, 15% were already distributing electronically—compared to 13% of the dailies in 1994. The flip side was really more significant—those with no plans to distribute electronically totaled 51% of the sample in 1994 among dailies; that number declined to 30% in 1995.

Steve Outing's list of U.S. newspapers with Web sites, compiled for *Editor & Publisher* magazine, was roughly 200 at the end of 1995—about 10% of all dailies, with a smattering of weeklies. This confirms our respondents' answers.

The needs of on-line publications will, in the near future, differ from exclusively print publications—they have room to publish more information, more points of view. And, they have more frequent deadlines. For magazines, this means a monthly may update its on-line offerings weekly or even daily. A daily newspaper's Web site may update its offerings (particularly for business, news and sports) two or three times a day now—and continuously in the near future.

Continuous updating may change the definition of a news embargo. The *San Jose Mercury News* Web site (Mercury Center) was the first to report on Jupiter-size planets orbiting two stars 35 light years away, even though the paper's science reporter, Glennda Chui, had not been able to wrangle a trip to the press conference in Austin, Texas. A day prior to the mid-January 1996 press conference, she had been faxed a press release by the PR department at San Francisco State University, where Geoffrey Marcy and David Butler had made their findings about the planets. Taking advantage of her friendship with Marcy, she interviewed him early on the morning of the press conference, then published her story on the Web just as his press conference began (and the embargo ended) a few hours later. The piece did not appear in the paper itself (the *Mercury News*) until the next day. Chui's "scoop" was discussed in *Science Writer,* the quarterly newsletter of the National Association of Science Writers.

Most newspapers with Web sites of their own do not yet allow the Web site to scoop the paper. Mercury Center was an early exception. In one sense, this episode is similar to a live radio or TV report from the site of the press conference. But there is a difference. The live broadcast is unfiltered by reporters. Chui filtered the Mercury Center report, but it still appeared at the same time as the event itself unfolded. Clearly, embargo rules may have to be redefined—and so do the operating rules of newspaper and magazine Web sites. The segment of a publication's readership that wants a certain type of news—astronomy in this case—will seek it out on the Web, ignoring the local, late, publication source. Thus, the local newspaper must "scoop itself" on its Web site or risk being scooped by someone else.

Today, few newspapers are doing much original reporting on their Web sites at all. In fact, in the name of avoiding union work-rule entanglements in a fluid, ever-changing, technology-based endeavor, newspapers have tended to set up their Web operations as distinct and separate entities, apart from the main newsgathering, newswriting activities. They have tended to hire HTML coders—and pay them more than reporters—even though coding is a skill that takes only a few weeks to learn.

Where there has been original reporting, it has been reporting that tends to exploit the nearly unlimited newshole of on-line publishing. We see school-by-school surveys, street-by-street stories, neighborhood-oriented advertising. It was easy to envision a "lawyer's version" of the O.J. Simpson story as it unfolded, or a feminist version, a Black version ... a Minnesota version. Unfortunately, most of the extra material tacked onto such a story theme may very well be archival, with little "added value"—rather boring for a typical reporter to assemble.

The next step, just beginning to happen, is the exploitation of on-line tools such as video and audio to make stories more vivid, and even fill-in-the-blank forms to personalize stories. What will the property tax increase mean exactly, to you as a reader and a property owner? What about changes in Social Security for someone at precisely your age, income, and state of residence?

It can be argued that such matters can be covered on paper, with sidebars and tables. But why would a reader accept an "approximate" answer from the local paper when he or she can have an answer that is exact, and when competitors are willing and able to provide that exact answer? Video and audio are, frankly, being overused on the Web—often provided for its novelty value. But as television and radio have proven again and again, the printed quotation is not always an adequate reflection of a source's tone or circumstance.

And what about new storytelling techniques? If I really wanted to explain New York City's budget history and budget priorities, for instance, perhaps I could do so by licensing a game such as Sim City, adding an appropriate set of New York City-like starting values, and allowing readers to download it from my Web site. Sim City is an interactive game that allows users to set policies for a model city of their own. The game reacts to the policies in a realistic way. In fact, much of the math behind it was developed by Jay Forrester at MIT in the 1960s. Few have read his book, *Urban Dynamics*. But millions have played Sim City.

Where do reporters and editors (or publishers, if these are to be new staff categories in the newsroom) get the new storytelling and math skills? And how will they manage the resulting organizations?

Many magazines are already there, as these surveys have shown. They tend to be smaller publications, however, with free (controlled) circulations. Their "print" business models require that 25% or more of revenue go into physical distribution and into circulation-building, mainly by direct-mail promotion. Those two expense categories are cut substantially by Web publishing. The "Web" business model should, in fact, be well suited to publications that are advertiser-supported, or that, like most newspapers, use subscription revenue to roughly cover or come close to covering manufacturing and distribution costs.

Smaller magazines have seen this already. They have tended to merge Web publishing with on-paper publishing staffs. They have also been more willing to explore new editorial techniques, using their uniqueness and editorial quality to preserve or gain a following. Magazine reporters and editors like to respond to our surveys by pointing out that their Web offerings are new publications in a new medium. The on-line product (whether it is *Columbia Journalism Review* or *Wired*) is not the same as the print product.

What will the reporter's contribution be, to this new age of math and games? On a magazine, the journalists often have specialized fiscal knowledge sufficient to handle the task. Newspapers may for a time have to resort to "new media syndicates" selling explanatory interactive features—and may be limited to stock, one-size-fits-all features instead of features customized precisely for their readers. One analogy in the newspaper business today would be the small paper that must use unedited or barely edited wire copy for national and international stories.

Weeklies and small dailies vary greatly in their sophistication when it comes to computer use; there is no clear pattern by location, circulation, or staff size. It appears instead that weeklies, like broadcast organizations discussed earlier, embrace new media technologies when they employ a "champion." The existence of that champion is somewhat random.

All this stands job-hunting advice for new media journalists on its head. The exciting, fun jobs for 1996 are with smaller news organizations, while major news organizations, particularly large dailies, are hiring but offer, on average, jobs that are more mundane for the present.

Not since the late 1960s, when advances in printing technology and list brokerage made small-circulation specialized magazines cost-effective, has there been such a revolution in the media business. The revolution in magazine publishing took a decade to unfold. The new media revolution is happening faster, but it will not occur overnight.

METHODOLOGY

The 1995 sample newspapers included all daily newspapers in the United States (including "Sunday-only" papers published with dailies of a different name), and several weeklies with a circulation greater than 2,500. The job titles among the 1,815 survey forms mailed to newspapers were evenly divided; 907 business editors and 908 managing editors. We saw no significant differences in responses from the two job titles. The usable responses represented publications with a total daily circulation of 23.8 million and Sunday circulation of 24.8 million—more than 40% of the nation's total newspaper circulation.

We also mailed questionnaires to 2,000 magazine managing editors in 1995. The magazine sample was not random; it was weighted toward business (but non-computer) publications; we tried to avoid small-circulation publications from narrow-interest groups. The sample did, however, include general membership publications of major professional organizations—particularly if those publications were produced by commercial publishers. Only three of the 479 usable magazine responses were from publications with a circulation less than 10,000. Total circulation of our respondents' magazines was 128.9 million—more than 10% of all magazine copies sold in a typical month in the United States. We consider magazines covering the computer industry as a special case; including them would have inflated the survey results on computer use by journalists.

Unlike the 1994 survey, we did not mail specifically to weeklies in 1995 (there were 1,500 "non-Sunday, nondailies" in the sample of 6,000 in 1994). Thus, when we compare newspapers in the 1995 sample to 1994, we specifically exclude weeklies in the 1994 sample to make the results comparable, where inclusions would have skewed the sample.

We also did not mail to freelancers as we had in 1994 (in 1994 we mailed to 1,500 freelancers who belonged to Investigative Reporters and Editors; this was roughly half the group's list). The response rate from freelancers had been low, and it is impossible to find a truly random sample to target with a survey.

Although we have been conducting informal surveys of broadcast or cable media for the past 2 years, we have not included broadcast or cable media in either survey except for some broadcast-oriented freelancers in 1994.

Focusing on dailies and magazines, providing more response time, and surveying mainly after Labor Day (the 1994 survey had been mailed early August) increased our response rate considerably, from 14% in 1994 to 22% overall in 1995. August is normally a good time to survey the media; in 1994 we wanted to survey before congressional election races heated up. But surveying in August 1995 would have caused us to run afoul of the August 24 U.S. introduction date for Windows 95 and the Microsoft Network on August 24. Survey forms mailed and returned in 1995 are summarized in the table:

	Number mailed	Usable responses	Incomplete responses	Returned as undeliverable	Response rate*
Magazines	2,000	479	17	128	26%
Newspapers	1,815	272	42	36	18
Total	3,815	751	59	164	22

* Total responses (complete and incomplete) divided by number mailed minus undeliverable.

For the sample as a whole, the results reported from respondents would be expected to be within plus or minus four percentage (+/- 4%) points of the responses expected from the entire universe of these categories of newspaper and magazine editors, 95% of the time—assuming that the sample is truly random. Here are the issues to consider with regard to the sample:

Nonresponse bias: The overall response rate for the 1995 survey, at 22%, is far above average for mailed surveys of this type. The response rate for magazines, at 26%, is outstanding. Furthermore, circulation histograms—the number of responses at each circulation range—display no obvious abnormalities.

Nevertheless, the newspaper responses represent a higher percentage of all newspaper circulation (40%) than they do of the sample to which surveys were sent (18%). Magazine responses, on the other hand, represent a lower percentage of all magazine circulation (more than 10%) than they do of the sample surveyed (26%).

Thus, we paid close attention to response variation by circulation size. (Cross-tabulation tables with circulation splits are reproduced in the original survey and are available on-line at http://www.mediasource.com, in the Extra-Extra section.) Except for questions on in-house libraries and production techniques, we saw no significant response variation by circulation size for newspapers, within limits of sample size, and only small variations for magazines. The "incomplete responses" noted in the table above lacked circulation data, so we could not match them to the overall circulation profile or to the 1994 responses. But we could use them to check for nonresponse bias.

We also offered respondents a copy of the complete survey results; they returned cards to us, certifying that they had completed the survey and sent it in separately. The 604 cards returned showed no obvious bias to one city or zip code.

Sample selection: While the newspaper sample both years was designed to cover all dailies and their associated Sunday editions (even when published with different staffs or under different names), the magazine sample was not "random," as discussed earlier. Compared to the magazine industry as a whole, the sample's circulation (the number of copies distributed, whether on the newsstand or by mail) is below average. But compared to subscriber-only publications, it is above average. The circulation frequency, however, mirrors the subscription-driven part of the industry.

There are about 12,000 magazines published in the United States that meet our criteria; we surveyed only 2,000. Even so, this represents the largest survey ever conducted of magazines in cyberspace.

We emphasize that the universe of responses from magazines in 1995 was slightly different than for 1994. We made no effort to keep to the 1994 magazine sample (several hundred titles had disappeared). And 1995 responses represent a slightly higher average circulation. Thus, the magazine sample is not completely

Frequency, times published per year	Number of magazines in sample
12	221
6	97
4	33
10	28
8	16
9	14
11	11
52	10
13	7
2	5
7	5
3	3
18	3
24	3
26	3
5	2
14	2
1	1
15	1
23	1
51	1
Total	**467**

random. Because the sample did not include computer-related magazines, however, the lack of randomness probably understates on-line use by magazines.

Subsampling: It should be obvious (but in published surveys often is not) that information from subsets of the overall sample is less reliable than for the sample as a whole. We provide complete cross-tabulations for our subsamples, at http://www.mediasource.com. In general, percentages from each individual cell in a cross-tab should be considered suspect when the cell size is less than 50—unless the percentage reported is very high or very low (above 80% or below 20%). Users who are interested in subsets of the data may wish to combine cells to get a valid cell size.

Questionnaire bias: We have minimized bias by exhaustively testing the survey; the fall 1995 version is similar in wording and layout to the 1994 version. In addition, there has been extensive and continuing presampling of non-random audiences (mainly at professional meetings of journalists).

The high response rate and low incidence of blank (unanswered) questions also suggest that the survey form was germane (asked the right questions) and understandable.

Timing: As the 1995 survey was being answered, Microsoft Network was just starting up, AOL was continuing its saturation marketing (a disk in every maga-zine, computer box, and—it seemed—anything that moved or stood still), and CompuServe was cutting its fees to match AOL. Prodigy and Delphi were trying to define a marketing strategy. The number of Web pages catalogued by the Lycos search engine grew from 8 million to 12 million (in 6 weeks!). In short, life was about normal in the on-line business. Since the survey was done, Microsoft's CEO Bill Gates has declared that the Internet looks like a good marketing bet, and that value-added services such as Microsoft Network must become more Internet integrated. The number of Web pages catalogued by Lycos grew to 60 million by August 1996.

Content Analysis in an Era of Interactive News: Assessing 21st Century Symbolic Environments

William Evans
Georgia State University

Content analysis is a set of techniques for systematically identifying message characteristics for the purposes of making inferences (often formal statistical inferences) about the contours of our symbolic environment. Content analysis has long been a staple of media research. Since 1965, content analysis has been used as the primary methodology in roughly one fifth of the articles published in mainstream media research journals (Cooper, Potter, & Dupagne, 1994; Potter, Cooper, & Dupagne, 1993; see also Moffett & Dominick, 1987; Wimmer & Haynes, 1978). In 1994 alone, more than 200 content analyses were published, and more than 70 of these dealt with popular news, assessing the coverage of topics such as crime (Barlow, Barlow, & Chiricos, 1994); foreign policy (Wells & King, 1994); military conflicts (Fico, Ku, & Soffin, 1994); biotechnology (Priest & Talbert, 1994) and AIDS (Gozenbach & Stevenson, 1994); and documenting the portrayal of women (Sage & Furst, 1994), African Americans (Lester, 1994), gay men and lesbians (Hallett & Cannella, 1994), and others.

Content analysis has played an important role in research programs that explore the relationship between news content and public opinion and behavior (e.g., see Fan, 1988; Noelle-Neumann, 1993; Protess & McCombs, 1992; Viswanath & Finnegan, 1995). Content analysis data continue to inform government policy regarding the diversity of ownership of news outlets.

Despite its popularity, however, traditional content analysis seems unsuited for assessing the role of news in the era of interactive media. Content analysis was developed in the 1940s and 1950s, an era in which news media were decidedly noninteractive and news outlets were relatively few. It was an era in which it was common for researchers and audiences alike to speak of "the news," a monolithic entity constituted by the handful of newspapers and radio stations that provided most of the news in most communities. It was a period in which readers and listeners had few choices except whether or not to attend to the relatively small number of news stories presented.

Content analysts in this era had little reason to question or to theorize the nature of content itself. News content was simply whatever was carried by the news media. News content was provided in discrete units called "stories" that were packaged in larger but also discrete units called "newscasts" or "newspaper sections." News consumers could be selective regarding which stories they attended to, of course, but they could not easily obtain access to a wide variety of news sources. In addition, consumers could not readily override the constraints of the isolated story as the basic news unit. They could not readily find or create links across stories or locate additional information related to the story to which they were attending.

This situation was convenient for media researchers, if not for consumers. Because news stories were fixed, discrete, isolated, and relatively few in number, researchers could gloss over or ignore potentially troublesome issues about the relationship between news content and audience consumption practices, issues that have become more salient and more troublesome in the era of interactive media, issues to which this chapter returns.

Another important aspect of early content analysis research was the assumption that the news audience was a mass audience that shared a relatively common culture. Researchers commonly presented data about news coverage of various issues and peoples without considering the characteristics of the journalists, news outlets, and audiences who, respectively, produced, disseminated, and consumed the news being studied. Content analysts assumed that the symbolic environment was relatively homogeneous and that the social context of news production and consumption was relatively unimportant.

Unfortunately, the methods and assumptions typical of content analysis have changed little since the 1950s. As we enter the era of electronic and interactive news, content analysis is ill-equipped to document and explain news content that is, in some sense, created by consumers in the process of being consumed. Since on-line news users may select from dozens of news sources, can explore links across news stories, and access text and video databases, it makes little sense to conceptualize news stories as the basic content unit. When each consumer can create his or her own path through on-line news resources, there may be as many news "stories" as there are consumers. In an era in which on-line news resources are often tailored to very specific political, occupational, racial, ethnic, gender, and lifestyle interests, it is no longer tenable to ignore the social context of news consumption.

In sum, content analysis as a method is unprepared for the challenges posed by interactive news. It is telling that there exists no published content analytic study of interactive news per se. In fact, it is reasonable to question the meaningfulness or even the possibility of doing content analysis in the era of user-centered hypermedia. This chapter describes the challenges involved in studying news in the era of interactive media and discusses several theoretical, methodological, and technological innovations that promise to help content analysts meet these challenges. I argue that content analysis is far from being irrelevant. Indeed, a reinvigorated content analysis will be crucial for understanding the new symbolic environments of the 21st century.

NEW MEDIA REALITIES

The proliferation of on-line news sources. The full texts of more than 110 daily newspapers are available on-line (Bjorner, 1995), as are the texts of more than 5,000 magazines, newsletters, and newswires (Orenstein, 1995). Moreover, these on-line news sources are no longer accessible only to those who establish accounts with the major on-line information providers such as LEXIS-NEXIS or Dialog. Increasingly, the full texts of print publications are being made available to consumers through commercial on-line services. For example, CompuServe's 2.7 million subscribers can search for and retrieve text from 33 daily newspapers for a flat rate of $24 an hour (albeit only during off-peak hours). In the next few years, consumers can expect to be offered easy and relatively affordable on-line access to a wide variety of print publications.

In addition to making the content of print publications more accessible, many traditional news outlets are developing news services designed specifically for the on-line environment. For example, more than 450 U.S. daily newspapers have developed sites on the World Wide Web (WWW) or are offering "electronic newspapers" through commercial on-line services. Interactive media initiatives are a high priority for all international news operations (e.g., CNN, Reuters) as well as for many local television stations. The joint projects of NBC and Microsoft are perhaps the most widely publicized recent examples of the inexorable convergence of the computer and news industries. The 21st century news consumer will have access to a wide variety of on-line news sources.

The proliferation of Internet newsgroups, mailing lists, and Web sites. There are several thousand Internet newsgroups and mailing lists, many of which serve as important sources of news for participants. Participants often post news stories they have found elsewhere on-line and wish to share with others on the list (Lewenstein, 1995), and these stories often become a focus of discussion among participants. (Such postings, at least on media-related lists, also inevitably occasion seemingly interminable discussions of the copyright issues involved in this practice.) Web sites offer user-friendly access to a wealth of news and information on many topics.

As one watches the Internet grow at an astonishing pace, it sometimes seems as if there is no political cause too radical, no conspiracy theory too implausible, no celebrity or television show too banal, no religious belief too unorthodox, and no sexual practice too exotic to warrant its own newsgroup, mailing list, or Web site. The Internet makes it possible for users to sample from a wide range of political and lifestyle orientations and ultimately to join a community of like-minded people. Moreover, Internet communities often play a role in determining what news outlets and stories participants will attend to and how they will frame and interpret the news. The Internet can cultivate immersion in virtual communities that are isolated from one another and from the mainstream culture. To the extent these communities serve as creators and mediators of news, our symbolic environment becomes more diverse (or at least more fractured) and, from the perspective of the content analyst, harder to document and to analyze.

The advent of hypernews. Few on-line news services take full advantage of the possibilities of hypermedia, but this will change in the near future as on-line news providers permit users to explore links to related texts, sounds, images, and archives. In this environment it is possible—indeed, likely—that no two users will pursue the same path and that no single user will pursue the same path twice. Therefore, it makes little sense for researchers to focus on the stand-alone news story as the basic content unit. Instead, researchers may be forced to reconceptualize content as being determined by the user-defined path through the news environment. In hypermedia news environments, it is not self-evident where the content lies. Studies of hypermedia will require fundamentally new ways of thinking about media content.

The advent of intelligent technology. Several providers of on-line news allow users to create their own on-line "newspaper," to specify the topics in which they they are interested (and uninterested), and even to customize the look and functionality of the newspaper. Electronic news clipping services scan the on-line news environment in search of stories that match the user's expressed interests. Intelligent software agents will soon be able to discern a user's news interests without direct user instruction and find and recommend news sources accordingly. Certainly, tools for searching the Internet will continue to become both more robust and user-friendly. And new tools for visualizing and organizing on-line information spaces will help users manage the torrent of available news and information (see Gershon & Eick, 1995; Nielsen, 1995; Rao et al., 1995). Content analysis becomes more problematic as more news consumers create their own, personalized news environments.

The emerging era of multiple news sources that are tailored by individuals to their own interests and accessed in hypermedia environments may seem daunting to the content analyst accustomed to fixed content and mass rather than interactive media. But all is not lost for content analysis, which, with proper attention to theory, a few methodological innovations, and the adoption and development of new computer tools, can provide potentially important data and insights regarding 21st-century symbolic environments.

MAKING SENSE OF DISPARATE BUT HOMOGENEOUS NEWS AUDIENCES

More news sources are now more readily available than ever before (at least for those who can afford on-line access to these sources), but this does not necessarily indicate that consumers of on-line news will choose to receive information from an ideologically wide range of sources. Rather, it seems likely that the power of inter-active media will be used by many consumers to narrow their news horizons, to sift information through very particular ideological filters. It seems unlikely, for exam-ple, that consumers will routinely ask automated news clipping services to find in-formation that challenges the consumer's most cherished beliefs. There is a need for research that explores the extent to which interactive news consumers will choose to narrow their news horizons, but the existing literature already documents a great deal of selective exposure and attention to media content. Reviewing this literature, Fiske and Taylor (1991) conclude that "most of us inhabit an environ-ment that is biased in favor of positions with which we already agree. . . . People tend to pick friends, magazines, and television shows that reinforce their own atti-tudes" (p. 469). Certainly, the many recently developed software systems for news filtering seem designed precisely to facilitate the narrowing of one's news horizons. These systems are touted by developers and users alike for their ability to provide users only with news that matches user interest profiles, protecting users from ex-posure to unwelcome news items (e.g., Muchmore, 1996; Smith, 1995). More than one dozen subscription-based news filtering services are now available. Such ser-vices invite users to specify news genres and topics in which they are interested. New filtering software then searches various news services for stories that match the user's interest profile. Users can readjust their interest profiles to insure that the computer returns just the right mix of stories. As the use of the term *filter* suggests, the features and interfaces offered with these tools facilitate the winnowing of is-sues and perspectives to which one is exposed. News filtering systems exploit and perhaps cultivate our tendency to selectively attend to news.

As we work to use news filtering tools and other software solutions to address the very real problem of information overload, we seem relatively unconcerned about the implications of information homogeneity. In the era of interactive news, the mass audience may be replaced by numerous audiences separated by great ideological distances but manifesting ideological homogeneity within groups. The Internet in particular seems to encourage the balkanization of news audiences. In-ternet newsgroups tend to be organized around specific political orientations, oc-cupations, and hobbies. Hundreds of "alt." newsgroups (e.g., alt.politics.perot, alt.religion.gnostic, alt.conspiracy.jfk) cultivate their members' (frequently self-proclaimed) alienation from mainstream culture and media. As the term *news-group* suggests, these communities are formed primarily to enable members to fo-cus on news and discussion regarding rather narrowly defined topics and issues. Moreover, there is little dialogue across newsgroups. Newsgroup members seldom interact across newsgroup boundaries, except perhaps to post purposefully incendiary messages (i.e., "flames"). Internet newsgroups provide powerful tools for linking the like-minded and insulating the like-minded from news about and conversation with the "unlike-minded."

Content analysts need to develop theories and methodological techniques appropriate for assessing the ideological distance between and within various groups of news consumers. As a start, content analysts might more rigorously model the relationship between news content and audience characteristics. Content analysts have been seemingly uninterested in modeling this relationship, perhaps because mass media audiences are relatively heterogeneous in comparison to many new media audiences. As we enter the era of narrowcasting, niche publishing, and Internet communities formed on the basis of very specific interests or belief systems, it becomes possible to more precisely match media content features to specific audience features. The proliferation of news audiences is, in some sense, a blessing in disguise in that it affords researchers newfound opportunities to explore the relationship between media content and social structure.

Just as the opportunity to advance media theory by assessing the diversity of online media content is great, so is the need to monitor our collective but fragmented symbolic environment. As Agre (1995) notes, conservatives have built their own electronic infrastructure to disseminate news and information, an infrastructure of which many liberals are barely aware. Similarly, to the extent that liberal views reach conservative audiences they increasingly do so as mediated by conservative gatekeepers such as Rush Limbaugh. On-line news and information sources and Internet discussion groups dedicated to specific political, ethnic, gender, and lifestyle interests may empower users, but, as noted above, they may also cultivate a willful, computer-facilitated isolationism. Content analysis can play an important role in monitoring the climate of opinion, identifying cultural tensions, and assessing intergroup relationships. By analyzing the information requested and consumed by various audiences, researchers can assess the extent to which new media are contributing to cultural balkanism, a type of isolationism manifested in audience preferences for news only about themselves and others like them. Content analysis can play a central role in research programs dedicated to understanding the social impact of new media.

INTEGRATING CONTENT AND USER DATA

As discussed above, the content of interactive news may be in large part user defined. When seeking news in a vast hypermedia database, users create a particular and perhaps idiosyncratic structure—or "story"—to adopt the once unproblematic but increasingly ill-fitted term. To identify content to be analyzed, researchers must know how users structure (and in some sense create) content as they explore on-line news resources. Fortunately, the same computers that so readily provide users with information also do a great job of collecting data about user access and browsing patterns. Moreover, computers can collect these data unobtrusively and in great quantities. Interactive news provides an opportunity for researchers to collect rich data regarding what users look at and for how long, how they get to a particular place or "story," how "deep" (in terms of pursuing embedded hypermedia links) they go, and where they go when they leave. Research on interactive news environments must necessarily blur the distinction between content and user analysis. But this blurring is potentially very productive in that it can encourage more sophisticated thinking about, and analysis of, the link between content and user behavior.

TOWARD AN EXPANDED METHODOLOGICAL TOOL KIT

The merger of content and user analysis will require that researchers master research designs and statistical techniques that focus on transitions and sequences. What hypermedia links are most commonly pursued by users? What is the nature of the content that users weave together as they browse on-line news? What content and what paths through the content are available but not utilized? To answer these questions, researchers may be required to rely on heretofore uncommon statistical tools such as lag-sequential analysis and graph-theoretical techniques. The latter can been used to map and categorize user browsing patterns, and the former can be used to discover which browsing patterns are most common (Gottman & Roy, 1990; Sanderson & Fisher, 1994). Also helpful will be approaches borrowed from the human-computer interaction literature, such as efforts to develop "grammars" of information seeking capable of characterizing the most common sequences and transitions utilized by users as they navigate information spaces (Olson, Herbsleb, & Reuter, 1994). Interactive news content is not static. Hypermedia and large information spaces shift some of the power to create content from the journalist to the news consumer. The new dynamics of news production and consumption require new methodological approaches to content analysis.

TECHNOLOGY TO THE RESCUE

The same computer and information technologies that are complicating the practice of content analysis also can be used to address these complications. The proliferation of on-line news resources provides an opportunity for researchers to search for and retrieve large quantities of texts from a wide range of news outlets, from small weekly newspapers to large metropolitan dailies. These resources facilitate larger, more comprehensive, and more powerful content analyses than were typical in the era of print indexes and microform archives. Content analysts must be wary of the indexing and archiving gaps, inconsistencies, and even inaccuracies typical of many on-line databases (Feola, 1994; Kaufman, Dyjers, &

Caldwell, 1993; Neuzil, 1994), but the broad scope and easy accessibility of these resources provide new opportunities for relatively cheap yet sophisticated content analysis projects.

Computer tools to manage both text databases (Dewire, 1994; Weitzman & Miles, 1995) and image databases (D'Alleyrand, 1992; Weiss, Duda, & Gifford, 1995) further facilitate content analysis, as do new tools for collecting, visualizing, and analyzing data regarding computer users' on-line activities (Eick, Nelson, & Schmidt, 1994). These tools provide on-line environments in which researchers can gather, store, retrieve, and manipulate materials for studies of news content and audience activity.

Software designed specifically to support content analysis is more readily available than ever before. Several increasingly sophisticated software packages for qualitative data analysis provide support for the on-line coding of texts and images (Richards & Richards, 1994; Weitzman & Miles, 1995). Software packages designed to support quantitative text and image analysis are also becoming more numerous and more sophisticated (Evans, in press-b; Roberts & Popping, 1993). And researchers have recently begun to examine in detail the reliability, validity, and cost-effectiveness of various computer-supported content analysis techniques (Evans, in press-a; Franzosi, 1995; Morris, 1994; Nacos et al., 1991). This literature can help provide a better sense of which methodological approaches and computer tools are most appropriate for various content analysis research questions.

The first expert systems for content analysis have been developed (Gottshalk & Bechtel, 1995; Hunter & D'Arcangelis, 1994), and Evans (in press-a) envisions a time in the near future when artificially intelligent software will not only largely automate content analysis data collection but also actively support theory development and hypothesis generation (see also Bainbridge et al., 1994). We must, of course, be wary of undue or unthinking reliance on computers, but it seems clear that computers will play a central role in the development and implementation of 21st-century content analysis research programs. Indeed, the complexities of new media content can perhaps be fully investigated and understood only with the help of computers.

CONCLUSION

Twenty-first century symbolic environments will be substantially, perhaps even fundamentally, different from 20th-century environments. Content analysis has a long tradition of helping us document and understand the role of media texts and images in our culture. This heritage will become more, rather than less, valuable as we try to understand journalism in an era of news proliferation and personalization. Content analysis can help us determine the contours, fault lines, and local topography of our numerous and often isolated symbolic environments.

Unfortunately, content analysts have been slow to respond to the new media realities, perhaps because the new realities are not yet well defined and because disciplinary change itself is sometimes painful—textbooks must be revised, new skill sets developed, and new paradigms embraced. Yet, these changes must occur if we are to insure the continued relevancy of education and research in communication and journalism. The emerging era of interactive news presents daunting challenges for media researchers, but it also provides opportunities to develop sophisticated research programs that can help us anticipate and understand the no doubt enormous influence of new media in 21st-century culture.

REFERENCES

Agre, P. (1995, August). While the left sleeps. *Wired*, 103–104.

Bainbridge, W. S., Brent, E. E., Carley, K. M., Heise, D. R., Macy, M. W., Markovsky, B., & Skvoretz, J. (1994). Artificial social intelligence. *Annual Review of Sociology, 20*, 407–436.

Barlow, M. H., Barlow, D. E., & Chiricos, T. G. (1994). Economic conditions and ideologies of crime in the media: A content analysis of crime news. *Crime and Delinquency, 41*, 3–19.

Bjorner, S. (Ed.). (1995). *Newspapers online*, (3rd ed.). Needham Heights, MA: BiblioData.

Cooper, R., Potter W. J., & Dupagne, M. (1994). A status report on methods used in mass communication research. *Journalism Educator, 48*, 54–61.

D'Alleyrand, M. D. (Ed.). (1992). *Handbook of image storage and retrieval systems*. New York: Van Nostrand Reinhold.

Dewire, D. T. (1994). *Text management*. New York: McGraw-Hill.

Eick, S. G., Nelson, M. C., & Schmidt, J. D. (1994). Graphical analysis of computer log files. *Communications of the ACM, 37*(12), 50–56.

Evans, W. (in press-a). Computer environments for content analysis: Reconceptualizing the roles of humans and computers. In O. G. Burton & T. Finnegan (Eds.), *Renaissance in social science computing*. Urbana: University of Illinois Press.

Evans, W. (in press-b). Computer-supported content analysis: Trends, tools, and techniques. *Social Science Computer Review*.

Fan, D. P. (1988). *Predictions of public opinion from the mass media: Computer content analysis and mathematical modeling*. Westport, CT: Greenwood Press.

Feola, C. J. (1994, July/August). The Nexis nightmare. *American Journalism Review*, 39–42.

Fico, F., Ku, L., & Soffin, S. (1994). Fairness, balance of newspaper coverage of U.S. in Gulf War. *Newspaper Research Journal, 15*, 30–43.

Fiske, S. T., & Taylor, S. E. (1991). *Social cognition* (2nd ed.). New York: McGraw-Hill.

Franzosi, R. (1995). Computer-assisted content analysis of newspapers: Can we make an expensive tool more efficient? *Quality and Quantity, 29*, 157–172.

Gershon, H., & Eick, S. G. (1995). Visualization's new tack: Making sense of information. *IEEE Spectrum, 32*(11), 38–56.

Gottman, J. M., & Roy, A. K. (1990). *Sequential analysis: A guide for behavioral researchers.* New York: Cambridge University Press.

Gottschalk, L. A., & Bechtel, R. J. (1995). Computerized measurement of the content analysis of natural language for use in biomedical and neuropsychiatric research. *Computer Methods and Programs in Biomedicine, 47,* 123–130.

Gozenbach, W. J., & Stevenson, R. L. (1994). Children with AIDS attending public school: An analysis of the spiral of silence. *Political Communication, 11,* 3–18.

Hallett, M. A., & Cannella, D. (1994). Gatekeeping through media format: Strategies of voice for the HIV-positive via human interest news formats and organizations. *Journal of Homosexuality, 26,* 111–134.

Hunter, D. C., & D'Arcangelis, R. M. (1994). A partitioned rule-based approach to content analysis. In D. Ross & D. Brink (Eds.), *Research in humanities computing* (Vol. 3, pp. 217–228). New York: Oxford University Press.

Kaufman, P. A., Dyjers. C. R., & Caldwell, C. (1993). Why going online for content analysis can reduce research reliability. *Journalism Quarterly, 70,* 824–832.

Lester, P. M. (1994). African-American photo coverage in four U.S. newspapers, 1937–1990. *Journalism Quarterly, 71,* 380–394.

Lewenstein, B. V. (1995). Do public electronic bulletin boards help create scientific knowledge? The cold fusion case. *Science, Technology, & Human Values, 20,* 123–149.

Moffett, E. A., & Dominick, J. R. (1987). Statistical analysis in the *Journal of Broadcasting,* 1970-85: An update. *Feedback, 28,* 13–16.

Morris, R. (1994). Computerized content analysis in management research: A demonstration of advantages and limitations. *Journal of Management, 20,* 903–931.

Muchmore, M. W. (1996, April 9). News you choose: Customizable news services. *PC Magazine,* pp. 199–222.

Nacos, B. L., Shapiro, R. Y., Young, J. T., Fan, D. P., Kjellstrand, T., & McCaa, C. (1991). Content analysis of news reports: Comparing human coding and computer-assisted methods. *Communication, 12,* 111–128.

Neuzil, M. (1994). Gambling with databases: A comparison of electronic searches and printed indices. *Newspaper Research Journal, 15,* 44–54.

Nielsen, J. (1995). *Multimedia and hypertext: The Internet and beyond.* Cambridge, MA: AP Professional.

Noelle-Neumann, E. (1993). *The spiral of silence* (2nd ed.). Chicago: University of Chicago Press.

Olson, G. M., Herbsleb, J. D., & Reuter, H. H. (1994). Characterizing the sequential structure of interactive behaviors through statistical and grammatical techniques. *Human-Computer Interaction, 9,* 427–472.

News and Information at the Crossroads: Making Sense of the New On-line Environment in the Context of Traditional Mass Communication Study

Kevin Kawamoto
University of Washington

INTRODUCTION

Many of the questions people ask about the Internet, the World Wide Web (WWW), on-line commercial database services and other emergent communication and information technologies and activities tap into a rich and diverse history of mass communication research. Mass communication has been around as a formal field of study for more than five decades in the United States. It has been studied under the purview of other disciplines for nearly a century, and goes back even further in Europe, to the work of French sociologist Gabriel Tarde, for example, who studied the relationship between newspapers and public opinion in the last decade of the 19th century (Lang & Lang, 1983, p. 128).

Over the years, mass communication research has had to grow with the successive invention, development and diffusion of new technologies: After books, newspapers and magazines came radio, broadcast television, cable television, and satellite transmissions. Even facsimile machines, when used in a certain way,[1] can be considered a mass medium. The field has had to expand the parameters of its research to accommodate the emergence of these technologies and their implica-

---

1. A growing number of organizations, such as *The New York Times,* the *Honolulu Star-Bulletin,* and the Interactive Services Association have experimented with sending news and information to subscribers via facsimile machines.

tions for society and for the existing mass media they often threaten to—but rarely do—supplant.

In the past decade or so, the field has been confronted with what may be its biggest challenge yet: what to do about the emergence of computer-based communication and information services. How do these technologies and services fit into the study of mass communication, and what kinds of questions should researchers be asking to make sense of the new on-line environment?

This chapter focuses on on-line news and information technologies. In the universe of possible aspects of new media to study (e.g., on-line entertainment, financial transactions, advertising, laws, and regulations, etc.), news and information seems (a) most relevant for the journalism profession; (b) most inclusive of the other aspects, if only tangentially; (c) large enough as a category to provide meaningful material for analysis; (d) diverse enough as a category to allow for different kinds of analysis; and (e) complex enough to work with on a practical, conceptual, theoretical and philosophical level. The chapter (a) describes the overall media environment and, in particular, news and information technologies and their implications; (b) constructs models for these new technologies and discuss a conceptual framework from which to view their development and impact; (c) shows the continuities and discontinuities of new media in light of the traditional mass media; and (d) suggests an immediate and long-term research agenda that combines a narrow focus on specific technologies and services with their larger social implications and potential consequences.

HISTORICAL CONTEXT

The mass communication researcher does not have to wander too far from his or her field to discover thoughtful questions to ask about emergent communication and information technologies: The history of traditional mass communication study and its collective body of research provide many insights and suggestions for studying the new media as well.

In recent years, Howard Rheingold has written and spoken about the formation of computer-mediated social groups, or "virtual communities," among people dispersed throughout the nation and world. "The question of community," he writes, "is central to realms beyond the abstract networks of CMC [computer-mediated communication] technology. Some commentators ... have focused on the need for rebuilding community in the face of America's loss of a sense of a social commons" (Rheingold, 1993, p. 12).

Although contemporary ruminations about the development of "communities" on the Internet are in vogue, questions about the relationship between mass media and community formation are about as old as the study of mass communication itself. In the 1920s, sociologist Robert E. Park was writing about the role of newspapers in helping to form a sense of community among urban residents who had emigrated to the city from rural areas or from other countries. Park believed that the press and journalists, whether they were conscious of it or not, tried to "reproduce, as far as possible, in the city the conditions of life in the village ... We are a nation of villagers." He and his colleagues at the University of Chicago saw communication as being synonymous with human connectedness. He wrote: "There is more than the verbal tie between the word common, community and communication" (Park & Burgess, 1924, p. 36). Their questions and ideas about the role of newspapers in helping to form a sense of community among people lacking close interpersonal interaction can help frame contemporary questions and concepts about the relationship between electronic communication networks and community-formation—not community in the physical or proximal sense, of course, but communities of consciousness. New forms of communication and information technology, some believe, are paving the way for a "global village," or what McLuhan has described as an "electrically configured world."

The name and ideas of Marshall McLuhan (1965; McLuhan & Fiore, 1967) are often invoked when the concept of an electronically mediated "global village" is discussed, but it is not clear exactly what McLuhan was envisioning when he coined the term. Although he used creative imagery in his descriptions, it is debatable whether he saw the level of interactivity that much of today's new media promise to deliver. If he was thinking of something akin to global television, such as satellite news networks, he was probably not far off target. Other concepts that are currently appearing in the literature, however, speak more directly to the centrality of human communication and interactivity at a global level and not just the dissemination of information across traditional borders. The idea of a "public sphere," a "global civil society," an "epistemic community," "cyberspace," and "virtual communities" suggests a kind of human cohesion based on common interests and concerns that transcends geographic and political boundaries and relies on emerging computer-driven interactive linking mechanisms.

Indeed, those who use and understand the emerging on-line environment are forming a kind of culture all their own. Attempts to describe this culture have given rise to the playful but telling use of popular metaphors such as "cyberspace," "the Net," "the Web," "Information Superhighway," "Infobahn," "Electronic Frontier," "cyberculture" and so forth. One can ride, surf, crawl, explore, browse, seek, search, fetch and lurk within and throughout intricately woven conduits of bits and bytes. In recent years, the complex and once virtually inaccessible world of computer networking has been collectively and consciously reduced to a kind of friendly, adventuresome place through the repeated use of these metaphors. The impetus to make computer technology and software "user-friendly," a term that is admittedly overused, has led to fundamental changes in many people's attitudes and behaviors in regard to personal computers. The de-professionalization of the language, concepts and applications of computer networking has allowed the computer to rapidly approach becoming a mass medium, although it still has a long way to go before achieving the high penetration levels of, say, television, telephone service, and even VCRs in American households. High-end computers and their various peripherals are not only relatively expensive compared to many other forms of personal technology, but they also require a certain degree of technical savvy on the part of the user and need to be perceived as serving some practical function before they can be more widely adopted.

Of course, technical terminology has not been eliminated as much as supplemented and, in some cases, subsumed by new terminology. Members of the computer-using cyberculture learn to "walk the same walk" and "talk the same talk," at least at a basic level, through initiation, imitation and immersion. They become familiar with the elements of their culture. A case in point: The September 1995 issue of *Wired* magazine contained a full-page ad that simply said, in three short lines:

> www.riddler.com
> This is your first clue.
> The rest is up to you.

If nothing else, this ad is a contemporary cultural artifact. It assumes a readership that can decode the message—a safe assumption considering the magazine—and act upon it. (In fact, the advertiser must be banking on this assumption.) But give the ad to people outside the "on-line culture" and they may not know what to make of it. They may not be able to "fill in the blanks," as people who share a culture are often able to do even without complete information, and realize that the full message is: "Go to a Web browser such as Netscape or Mosaic, type in the uniform resource locator (URL) in the appropriate location, preceding it with 'http://' if necessary, and see what a great Web site this is!" URLs seem to be everywhere these days—in newspapers, magazines, on television, posters, flyers, and even on the sides of buses, an indication that the on-line culture comprises a potential mass market.

Culture and language are closely tied together. Writings in or about the new media are replete with their own peculiar language—metaphors, acronyms, signs, symbols, emoticons and URLs, as well as a host of delineations that were never quite as important as they are now, such as "in real time" versus asynchronous time and "real world" versus cyberspace. Old words such as "bulletin boards," "links," "site," and "icon" have taken on new meanings or dimensions. New terms have been made up and widely adopted. A code of Internet behavior has evolved and is passed on, for example in on-line discussion groups, from "old-timers" to newcomers, in the same way that culture, in the *non*-on-line world, is "transmitted [from] generation to generation, with the responsibility given to parents, teachers, religious leaders, and other respected elders in a community" (Brislin 1993, p. 23). Although all people, in whatever "world" they live, may not adhere to these codes, that doesn't invalidate the concept of an evolved sense of manners, or what might be referred to in cyberspace as "netiquette."

The researcher interested in studying or understanding these on-line cultures, and having a vocabulary to describe and discuss what they observe and learn, should look at the large body of work in cross-cultural and intercultural communications. People like Brislin, Gudykunst, Hofstede, Hui, and Triandis and others[2] can shed light on the more obscure elements of culture that are difficult to define and reflect upon. Key terms and ideas from this literature can be applied, when appropriate, to on-line cultures and activities. Reading about the methods of intercultural or cross-cultural communication research also alerts the researcher to the numerous issues that must be considered when studying diverse people.

Virtual communities are an important phenomenon because they, in some ways, confound the traditional mass communication process and, in other ways, extend it. Information can travel along nontraditional routes in sometimes circuitous ways—the way information might spread through informal interpersonal channels in a village—and this poses legitimate concerns about accuracy and reliability. On the one hand, professionalization can more easily be compromised; on the other, a wider integration of content and perspectives can be accommodated. The effects of emerging new channels of news and information dissemination and exchange, and the communication processes involved, have yet to be thoroughly examined.

Concerns about the "effects" of new media such as the Internet have been increasingly voiced by parents, politicians, law enforcement officials, educators, mental health professionals and nonprofit and special interest groups. Others extol new media's potential for information liberation and for facilitating the emergence of a "global civil society" where all voices can be heard and governance is communitarian. This concern about effects—good, bad, or indifferent—is the lifeblood of mass communication research. Traditional mass communication study is largely built on "effects" research, from Lazarsfeld's "two-step flow" to Rogers' "diffusion of innovations"; from Noelle-Neumann's "spiral of silence" to Shaw and McComb's "agenda-setting function." Not to mention all the propaganda and public information research conducted during World War II. The question of whether media messages have powerful, not-so-powerful or mediated effects on audiences—and the specific nature of those effects, if any—is central to any comprehensive mass communications textbook. The answer to that question has often generated intense debate in various sectors of society and has been the basis for public policy formulation.

Thinking about effects or outcomes on a much larger scale has also become part of the mass communication research tradition. The Frankfurt School and its intellectual descendants and offshoots, often referred to as critical scholars, have long been concerned about the ideological and economic power wielded by large media institutions, a concern that can only intensify as "media giants" continue to get

---

2. See bibliography in Brislin (1993) for an extensive list of books and articles by cross-cultural studies scholars.

bigger through mergers and acquisitions and involve themselves in expanding media enterprises. International communication has been examined from a wide range of perspectives, including questions about the patterns of international information flow, national image, comparative media systems, media ownership and the consequences of media conglomeration.

Another obvious area of mutual interest between traditional mass communication and new media studies is the application of First Amendment law, free speech, privacy and intellectual property issues. The question of what's fit to print on-line has been asked in courtrooms, classrooms, boardrooms, and legislative offices everywhere. It will continue to provide considerable fodder for future discussion and debate. For more than a decade, Ithiel de Sola Pool and others have argued that the electronic media should be treated analogous to the print media, i.e., unhindered by government regulation. "Electronic media," he writes, "as they are coming to be, are dispersed in use and abundant in supply. They allow for more knowledge, easier access, and freer speech" (Pool, 1983, p. 251). Others like Vincent Moscow (1989) are more wary of the free market as a regulator of new information technologies and are concerned that media monopolies left unregulated threaten to dismiss the public interest in the pursuit of profits. Still others like Majid Tehranian (1990) advocate the combination of local control and global outlook—"think globally, act locally"—as part of a communitarian approach to new media opportunities. These perspectives represent at least three major strains in public policy discourse and are closely tied to particular ideological influences in a global political economy.

On a more "nuts and bolts" level, Lance Rose (1995) has compiled an entire book devoted to an emerging body of law and policies governing the on-line world, an area of legal study very much in its infancy stages. To regulate or not to regulate is hardly a new question, however. As far back as 1644, John Milton's *Areopagitica*, a classic piece arguing against government-regulated printing licenses in England, forcefully and eloquently put forth reasons not to control the press out of the ostensible fear that lies would get mixed up with the truth. "Let [Truth] and Falsehood grapple..." he encouraged.

Milton's philosophy, at least in theory, was that "Truth" would percolate upwards out of the muck of diverse ideas and would ultimately be victorious. Mass communication, however, whether espousing truth or falsity, has often posed a challenge to authority. Throughout history, many similar concerns, fears and motivations have driven attempts to regulate and censor expression through a mass medium. Confronting these attempts have been a series of counterforces, resulting in a tenuous and delicate balance in a complex struggle for power, control, empowerment and liberation both among and within government, the marketplace and civil society. Current activities and discussions surrounding the regulation of the Internet and other new communication and information technologies are part of this cyclical struggle to define and enforce the boundaries of expression. In the context of this historical accumulation of intellectual and political discourse, it seems reasonable that today's "grappling" with the issues concerning free speech and new media can have important and serious consequences for future policy making, as past ideas, such as Milton's, continue to have impact and relevance to the present.

In reviewing the history of traditional mass communication research and comparing it with concerns about the new media, one thing becomes clear: Many of the same questions once asked about the traditional media are now being asked about the new media. The technologies may be different, but many of the concerns have carried over. In this sense, the rise of new media is not so much a revolution as an evolution in the mass communication process. The difference may be semantic but is crucial: "Revolution" suggests an end and a beginning; "evolution" assumes a progression, a continuation. To dwell too singularly on the revolutionary aspects of new media may suggest a sudden discontinuity in the evolution of mass communication. In reality, both continuities and discontinuities are apparent. As mass communication study advances into the future, the emergence of a new media environment should be a part of its evolution, like an organism that has transmuted over time, maintaining key similarities while sprouting important differences for adapting to its surroundings. Roger Fidler has suggested the term "mediamorphosis."[3] While Fidler uses the term in a positive sense, others see the transformation as being more Kafka-like, potentially harmful and undesirable. This sense of foreboding about the consequences of technological change is recurrent in history, popular literature and science fiction.

Movies from the past and present reflect this foreboding. For example, Dr. Morbius, in the 1956 film *The Forbidden Planet*, is unaware that his own jealous, evil "id" is powering the murderous supercomputer hidden deep within the planet Altair. The movie was released at a time when computers, then colossal and intimidating to most people, were being developed for wider application in society. In recent years, there has been an unleashing of movies about computers gone out of control or being used for criminal intent or having the potential for mass destruction.[4] The Frankenstein genre (i.e., the notion of the well-intentioned scientist inadvertently creating a monster) has translated well to films involving computers that play on age-old fears and uncertainties about where technology will take society (or, perhaps more frightening, where society will take technology). While in-

---

3. Roger Fidler developed *The Tablet*, a flat-panel display newspaper whose research and development was funded by Knight-Ridder before the project was discontinued.

4. Aside from *The Forbidden Planet,* some of the other early films about sinister computers, sinister people using computers or computers gone out of control include *Desk Set* (1957), *2001: A Space Odyssey* (1968), and *Demon Seed* (1977). In the 1980s there was *Tron* (1982), *War Games* (1983), *Electric Dreams* (1984), and *Terminal Entry* (1986). And thus far in the 1990s, there has been *Sneakers* (1992), *The Lawnmower Man* (1992), *Ghost in the Machine* (1993), *The Double 0 Kid* (1993), *Brainscan* (1994), *Hackers* (1995), *The Net* (1995), *Johnny Mnemonic* (1995), and *Copycat* (1995).

dividually most of these movies are simply entertaining, as a whole they tend to raise the specter, however subtle and subconscious, of "What if...?" The student of popular culture could do a fascinating analysis of the common themes that emerge from these films and suggest social, psychological, and historical factors that may contribute to their production and popularity.

In the real life world of journalism, technology and international relations, according to Johanna Neuman, particular cycles from one era to the next are apparent. "The changes in international relations brought by the satellite and the computer, by digital technology and global networks, by CNN and real-time television, are profound," she writes. "These changes are marvelous and sobering and frightening and dramatic, but what my readings through history demonstrated is that they are not new" (Neuman, 1995, p. 7). What is new, she allows, is "the speed with which technology is assaulting the political world." Nevertheless, she continues, the "changes may be coming faster now, but every excursion into history confirms a consistent pattern of social change. Whenever a new communications technology arrived on the scene, diplomats scoffed at the new invention, journalists boasted that their influence had exploded, the public noticed that the world was shrinking, as if the boundaries of home were stretching to meet the horizon" (1995, pp. 7-8).

Mass communication research surrounding traditional media, then, can clearly help put the new media environment into context and help raise interesting questions for further study. This is not to claim, however, that there is nothing new about new media, because there is. The next section raises a number of potential topic areas in both the structure and process of new media that could benefit from thoughtful and creative elucidation.

THE FUTURE OF NEWS AND INFORMATION

An extensive review of the new media literature (books, articles [newspapers, magazines, scholarly journals], proceedings, and electronic databases) reveals a confluence of factors, all interrelated and sometimes overlapping, that have contributed to the rise in new media. They are summarized here as:

- *Digitization:* Text, sound, graphics, video, and other information can be converted to binary digits and stored, retrieved, transferred, and otherwise manipulated by computers and their users.
- *Personal Computers:* User-friendly and relatively affordable personal computers grew rapidly in the 1980s. Computing power has steadily risen as prices and size have fallen.
- *Software:* A vast array of software for people of all ages and for a wide variety of purposes has been developed, much of which was intentionally created for the nonexpert.

- *Networking Capabilities:* Hardware and software that allow computers to communicate with each other via local area networks, wide area networks and through the Internet has been developed.
- *Networking Infrastructure:* A national and global information infrastructure based on electronic networks, computers and "protocols" that permit interoperability is being built. A federal policy initiative to encourage development of such an infrastructure was given impetus by the Clinton-Gore Administration in 1992.
- *Policy Formulation:* Federal, state and local initiatives are addressing the changing communications environment by re-examining laws and regulations that impact communications industries, services and activities. Most famous is the Clinton-Gore September 1992 report on the National Information Initiative. Also, major discussions took place in Congress in 1994 and 1995 to revise the nation's communications laws, which ultimately resulted in The Telecommunications Act of 1996, signed by President Clinton in February 1996. The Act was a major overhaul of the nation's communication law.
- *Corporate Consolidations:* Discussions or occurrences of mergers, takeovers and joint-ventures involving both similar and different kinds of companies (e.g., one software company merging with another software company as opposed to a telephone company merging with a cable company) have been taking place and are expected to increase as a result of The Telecommunications Act of 1996.
- *Technological Convergence:* The idea that previously discrete technologies are "coming together" is being realized through digitization and technological innovation. For example, it will be possible to receive telephone service through the cable television system or for a telephone company to offer video-on-demand. This is related to digitization, corporate consolidations and broader bandwidth, as well as the ability to compress and decompress digital signals.
- *Broader Bandwidth:* For a ubiquitous communications system in the home, office, school or business, there must be large carrying capacity in the lines of transmission. The deployment of coaxial cable and fiber optics permit more information to reach their destination in acceptable quality, but telephone companies are also increasing bandwidth through Integrated Services Digital Network and Asymmetrical Digital Subscriber Line. Related to this are technologies that compress and decompress digital signals to allow for more information to travel through set conduits.
- *Computers in the Workplace, Education, Home, and Elsewhere:* Computers and computing technology are proliferating in offices, stores, banks, schools, libraries, fast-food restaurants, health-care facilities, and in the home. Sophisticated computers are sometimes "disguised" as cash regis-

ters, on-line catalogs, automatic teller machines, and security systems, but
they are all electronic and work with digital information.

- *Entertainment:* Interactive CD-ROMs, special effects on film and television, computer-generated animation, video-on-demand, computer and video games, certain kinds of electronic discussion groups, and other forms of
computer-based entertainment or tools for creating entertainment are being
developed and used.

- *News and Information:* There has been an increase of on-line news and information. (See below.)

This confluence of factors has given rise to the "new media," a term which encompasses everything from the Internet and World Wide Web to LEXIS-NEXIS
and the Bloomberg on-line information services. John Pavlik,[5] author of a textbook on new media, includes within its scope "electronic media technologies that,
driven by computers, are rapidly merging into a single digital communication environment." He raises an important cautionary note: The nature and impact of
these technologies are "largely unsettled, primarily because they are still developing both in their form and function" (Pavlik, 1995, p. xiv).

New media, then, are electronic, are driven by computer technology, involve
some aspect of interactivity, reach a mass audience or are accessible by a mass audience (although not necessarily simultaneously, like television and radio) through
electronic networks, and are differentiated from traditional mass media in a number of ways including production, distribution, display, and storage. Those who
study the new media, just like those who studied the traditional mass media in the
early days of the field of mass communications, are an interdisciplinary lot. They
come from art, engineering, computer science, journalism and mass communication, health sciences, law, business, and so forth. Likewise, the scholarly roots of
traditional mass communication study can be traced to psychology, advertising
and propaganda, public opinion, sociology, engineering, rhetoric and speech, film,
journalism, political science, philosophy, literary criticism, and education.

The scholarship in mass communication is abundant but difficult to categorize
neatly, in large part due to the assimilated nature of the field's origins. Despite the
vast range of topics that could fall under the rubric of "mass communication," they
all share at least two common denominators. The first is the mass media. Books,
magazines, film, radio, and television (both cable and broadcast) are the typical
"channels" of mass communication. The second common denominator is the relationship between the mass media and society (i.e., individuals, groups, audiences, cultures, nations and so forth.)

The model of the traditional mass media, however, has to be reconstructed for
new media. Bittner (1986, p. 12) provides a typical textbook definition of mass

---

5. Pavlik is director and professor of the Columbia University Center for New Media.

communication as "messages communicated through a mass medium to a large number of people." The general idea of messages or content generated from a single source and dispersed to many reception points is central to understanding how mass communication works at its simplest level. DeFleur and Dennis (1991, p. 20) provide a more explicit description of the mass communication process by identifying five distinct stages in the process:

1. A message is formulated by professional communicators;
2. The message is sent out in a relatively rapid and continuous way through the use of media (usually print, film, or broadcasting);
3. The message reaches relatively large and diverse (i.e., mass) audiences who attend to the media in selective ways;
4. Individual members of the audience construct interpretations of the message in such a way that they experience subjective meanings, which are at least to some degree parallel to those intended by the professional communicators;
5. As a result of experiencing these meanings, members of the audience are influenced in some way in their feelings, thoughts, or actions; that is, the communication has some effect.

Emergent news and information technologies, however, do not necessarily adhere to traditional models of the mass communication process. In traditional ways of delivering news, the receiver of news plays a more passive (but not entirely inactive) role, allowing "professionals" to survey the universe, select what is worth attending to, write about it and present this information in the form of a newspaper article or broadcast report. McQuail and Windahl (1986) have compiled a useful compendium of communication models beginning with Lasswell's (1948) "Who says what in which channel to whom with what effect?" formula, Shannon and Weaver's (1949) "mathematical model" and continuing from there with more complicated revisions, modifications, elaborations, and new creations by others. As these models evolve, one simple fact becomes clear: Communication is complex. It is at once a process, a structure, a goal, a behavior, a symbol, a function, a tool, and many other things. These models will evolve further when applied to new media.

The new media assume a more active consumer of news and information, someone who has to, first and foremost, learn how to use a particular technology and then search for information or in some way negotiate a method of selection and delivery beyond turning on a television, subscribing to a newspaper or picking up a magazine at the newsstand.

The way people regard and engage with on-line news and information has only recently been the subject of study. Unfortunately, trying to describe on-line news and information is like trying to describe food. That one is electronic and the other is to be eaten are correct but not particularly helpful. Forming conceptual categories of food (meat, starch, dairy, vegetable, and fruit) is the first step in discussing food's heterogeneity. Likewise, the connoisseur of on-line news and information knows that the technologies and services now available for consumption would be difficult to discuss without first creating at least broad organizing categories. These categories are not necessarily distinct from each other, and they will not be minutely described here because they are defined thoroughly elsewhere. They are raised for the purposes of further discussion.

THE INTERNET AND WORLD WIDE WEB MODEL.   This is probably the most complex of the categories because it really is not a model in itself but contains a number of models within a large, overarching model. Essentially, the Internet and WWW link the user to news and information—most of it free, some of it free with registration, some of it with a price. Home pages on the Www have been erected by the traditional news media as well as other organizations and individuals.

NEWSGROUPS AND LISTSERVS.   Newsgroups can be online discussions or read-only information built around subjects of interest. Listservs are subscribed to and can be like electronic newsletters (no discussion involved) or discussion groups with or without a moderator.

COMMERCIAL INTERNET ACCESS PROVIDERS.   America Online, Compuserve, Prodigy, Delphi, and the Microsoft Network, as well as smaller companies, not only provide access to the Internet but also offer news and information from traditional news media, their own news division, or a combination of both. Many sponsor their own on-line discussion groups.

COMMERCIAL DATABASES.   LEXIS-NEXIS, Mead Data, Westlaw, and a host of other companies feature extensive databases that can be accessed on-line for a cost.

NONCOMMERCIAL DATABASES.   Universities and government sites, nonprofit groups and other organizations allow on-line access to databases without cost.

TRADITIONAL WIRE SERVICES.   The traditional wire services such as Associated Press, Reuters, and the United Press International send articles on-line to subscribers.

SUBSCRIPTION NEWS.   These come in several subcategories based on cost and content. Bloomberg, Dow Jones, and Reuters each offer on-line business and financial news. This service is relatively expensive. (In 1996, for example, Bloomberg charged more than $1,000 per month to lease its control box and monitors.) General on-line news from particular newspapers is available through bulletin board services or linkaging enterprises such as AT&T Interchange. Costs vary. (As mentioned earlier, however, many news organizations offer news and information online at no cost on the WWW.)

These models represent an overwhelming amount of news and information that is produced, distributed, displayed and stored in different ways. As such, it is difficult to generalize about on-line news and information without first attempting to categorize them by general characteristics. The first stage of DeFleur and Dennis' description of the mass communication process (i.e., "A message is formulated by professional communicators") is met by some, but certainly not all, of the on-line news and information services. Many well-established and widely respected news media (both print and broadcast) are putting their content on-line either for free or at cost. The veracity of the content is, to some extent, dependent on the name credibility of the source, not unlike the way people regard news and information that are not on-line. Certain news media have established credibility in the mind of consumers; others have not. The benefit to existing traditional media going on-line is that they can capitalize on their credibility, whereas on-line "start-ups" have to build up their credibility and have a number of other obstacles related to their newsgathering and delivery resources.

There are on-line news and information services that are not formulated by professional communicators (e.g., trained journalists and editors) and do not necessarily reach a large and diverse audience. Professional news organizations may be suspicious of their nonprofessional counterparts on the Internet and WWW, but there is probably a place for both kinds of approaches, especially considering the self-governing, decentralized and somewhat anti-establishment nature of the Internet once it became a popular networking tool.

The Internet is a virtually boundless receptacle of information, which can be its best and worst characteristic. The tradeoff for quantity may have to be a certain degree of quality. Ideally, if anyone can be an electronic publisher on the Internet, that means professional and nonprofessional, rich and poor, powerful and powerless, educated and uneducated, conservative and liberal, civil and uncivil can potentially have a voice. In reality, for many reasons, access is not equal, but nevertheless there are many perspectives represented in the universe of on-line news and information, and the playing field is at least more level than with many forms of traditional mass media. Consequently, the Internet, WWW and other forms of computer-based news and information services have become flooded with content.

Some believe that this excess will insure the survival of traditional mass media functions because journalists and editors will be needed to survey the morass of information and cull from it the things most worth attending to by the public. Moreover, trained professionals are supposed to be critical about the information they gather—checking for accuracy, veracity, and logic. Now more than ever, the argument goes, information needs to be filtered. This argument may be true in the short-term, but the solution to "information overload" is not necessarily always going to be filtering in the traditional sense. The growth of search engines on the Internet, WWW and large databases (including on individual home pages) suggests another solution. Efficient, logical, and purposeful searching skills are becoming mandatory for new media users if they ever hope to find information of interest and use to them.

The ability to use search engines and to perform sophisticated search strategies are a combination of mechanical and conceptual skills. As to the question of quality and credibility, new media literacy skills need to be considered. The ability to critically evaluate information in itself and in the larger context of where it comes from is important regardless of how or where the information is found. Perhaps conventional definitions of "quality" and how it is achieved need to be questioned or expanded. The very definitions of "news" and "newsgathering" may have to be rethought in light of new media as professional standards commingle with what might be described as amateur and grassroots reporting initiatives. But the presence of the latter should not be exaggerated. In many communities and villages there are both formal, professional channels of newsgathering and dissemination as well as informal, nonprofessional channels. Both co-exist; both have their detractors. The important thing about new media is that it can be used as a professional tool by traditional news and information providers, allowing these providers to add value, flexibility, and creativity to what they already offer.

The fundamental differences between the traditional mass media and the new media have to do with space, time and choice. New media are not confined by space or time restrictions the way traditional media are. The goal, taken from the ad of one new media news prototype, is not to fill a newshole but to aim for completeness and context. This goal, which could actually apply to a variety of different new media news prototypes, is possible through the use of hypertext links and other mechanisms for immediately accessing related information. For example, if a person is reading an article on Bosnia in a new media format and wants more information, that person could retrieve detailed background information about Bosnia or the region, a map showing where Bosnia is, related current articles, related archived articles, and even moving sound and video, all at the click of a button. Some new media services will also allow the user to enter an on-line discussion about various topics covered in the electronic publication. An article on migraine headaches, for example, could provide a link to an on-line discussion group for people who suffer from migraine headaches. They can commiserate with each other, learn how others cope, discuss medications and new treatments, and so forth.

In some ways, more information is more reliable because a more complete picture is possible. Writing for the new media often places more responsibility on both the consumer and the news provider. Traditionally, news has not been footnoted, although it is supposed to be sourced. With new media, details and quotes can be sourced as well as footnoted, hypertext links can be added to related articles and topics, and readers can interact with the writer or the respective news organization through electronic mail to complain, offer suggestions, make comments, or ask questions. The news provider can place stories in larger context with links to both current and archived information but without the same restraints imposed by space and time as exists with traditional news media. Information can be updated, modified, and corrected quickly. The potential for reader or viewer feedback exists with traditional media as well, of course, through letters to the editor or phone calls, but the convenience, speed and ease of interaction tend to be greater, generally speaking, with the new media.

More problematic for some people is the notion of customized news—giving the user a choice about what kind of news and information he or she wants to get. If a user only wants information about a highly publicized trial, for example, that option might be programmed into the new media service. On the other hand, if a user does not want any information about a highly publicized trial, that option might also be programmed.

There is an obvious good and bad side to customized news. Should people be able to choose the news they want to read? Or is that the role of professionals who "know better"? One could argue that people interact with the traditional media in similar ways anyway, attending only to information that interests them and ignoring the rest. The counterargument is that the mere exposure to many different news items at least offers the possibility of attention, whereas blocking out exposure altogether precludes the possibility of attention. This is a complicated issue. Is it elitist to believe that professional journalists and editors know what the public should be concerned about? Or is the public capable of setting its own agenda—choosing what it wants, when it wants it, in as little or as much amount as it wants?

Questions about the responsibility of the news media to the public, to building an informed electorate and to maintaining a free and open democracy are as relevant to the new media as they ever were in history. The easy answer to whether customized news is good or bad for a person is that it depends on the person. Whether it is good for society is a more difficult question to answer, but a good way of beginning to think about this issue is to examine the role of a free press in society, a topic on which there is no shortage of literature in the study of mass communication history.

The serious study of the new media presents many exciting challenges and opportunities for the mass communication researcher to break new ground. Like new media themselves, the study is convergent, bringing together people from many fields and disciplines, not unlike the early days of mass communication study with the political scientist Lasswell, the mathematician-sociologist Lazarsfeld, the social psychologist Lewin and the clinical psychologist Hovland. Schramm was an English professor. Every field has its forerunners and innovative thinkers. Who are those people in new media today? Who are the major technology writers? What are the major research institutes and programs? Where are the innovative projects and programs—well-known and lesser known? What publications should people be reading to keep up with the technology? What kind of research is being conducted?

New media study will have as many components as traditional mass media study, if not more. To truly understand the emergent on-line environment, there must be a multifaceted research approach that brings together the study of history, effects (at many different levels), government and policy, economics, international and cross-cultural aspects, technology and engineering, law, journalism, information sciences, and more. The study of mass communication is one of the few fields that already encompasses this vast body of research interests. The opportunity now is to set the research agenda for the new media—to begin collecting, categorizing, classifying, building and refining models, experimenting, developing theories, comparing and contrasting, following policy and considering implications and consequences at every conceivable level in society. Now is the time to look back, and move forward.

REFERENCES

Bittner, J. R. (1986). *Mass communication: An introduction* (4th ed.) Englewood Cliffs, NJ: Prentice-Hall.

Brislin, R. (1993). *Understanding culture's influence on behavior.* Orlando, FL: Harcourt Brace & Co.

DeFleur, M. L. and Dennis, E. E. (1991) *Understanding mass communication* (4th ed.) Boston: Houghton Mifflin.

Lang, K., & Lang, G. E. (1983). The "new" rhetoric of mass communication research: A longer view. *Journal of Communication, 33*, 128-140.

McLuhan, M. (1965). *Understanding media: The extensions of man.* New York: McGraw-Hill.

McLuhan, M. & Fiore, Q. (1967). *The medium is the massage.* New York: Bantam.

McQuail, D., & Windahl, S. (1986). *Communication models for the study of mass communications.* London: Longman.

Moscow, V. (1989). *The pay-per society: Computers and communication in the information age.* Norwood, NJ: Ablex.

Neuman, J. (1995). *Lights, camera, war.* New York: St. Martin's Press.

Park, R. E. & Burgess, E. W. (1924). *Introduction to the science of sociology.* Chicago: University of Chicago Press.

Pavlik, J. V. (1996). *New media and the information superhighway.* Needham Heights, MA: Allyn & Bacon.

Pool, I. d. S. (1983). *Technologies of freedom.* Cambridge, MA: Belknap Press.

Rheingold, H. (1993). *The virtual community: Homesteading on the electronic frontier.* Reading, MA: Addison-Wesley.

Rose, L. (1995). *NETLAW: Your rights in the online world.* Berkeley, CA: McGraw-Hill.

Tehranian, M. (1990). *Technologies of power: Information machines and democratic prospects.* Norwood, NJ: Ablex.

About the Authors

Patricia Aufderheide (paufder@american.edu) is an associate professor of communication at American University and is a senior editor at *In These Times* newspaper. A prolific journalist, she has written extensively on democratic applications of communications and has worked as a policy advocate for the United Church of Christ.

Diane L. Borden (dlborden@gmu.edu) is an assistant professor of communication at George Mason University. She holds a PhD in communications from the University of Washington. She came to academe after a lengthy career in professional journalism during which she worked as a newspaper editor and publisher in several cities. She has a keen interest in how the media and other cultural institutions—including the judicial system—historically have constructed social reality, particularly images of women and minorities. Her research focuses on the intersections of communication, gender, and the law.

L. Carol Christopher (cchristo@weber.ucsd.edu) is a PhD candidate in the Department of Communication at the University of California, San Diego. Her research interests include newspapers, journalism, technology, work organization, and feminist theories and methodologies. Christopher returned to academia after nearly 20 years in the newspaper industry, where she was a newsroom systems analyst and electronic publishing consultant, and a systems editor at *The Dallas Morning News* and the *Denver Post*.

William Evans (evans@gsu.edu) is an assistant professor of communication at Georgia State University, where he teaches courses in communication theory, communication technology, and visual communication. He holds a Ph.D. from Temple University. His research interests include computer-supported content analysis, interactive media, and science and health communication. His research has been widely published.

Jan Fernback (fernback@rintintin.colorado.edu) is a doctoral candidate in the School of Journalism and Mass Communication at the University of Colorado at Boulder. She is currently writing a dissertation on computer-mediated social relations, which assesses the concept of community as manifested in cyberspace. Her research interests include utopianism and new media technologies, and the anthropology of cyberculture.

Lisa St. Clair ("Kerric") Harvey received her PhD in communications from the University of Washington. Currently a member of the School of Media and Public Affairs faculty at The George Washington University, she teaches courses on innovations in electronic media, cultural studies, aesthetic theory, research methods, and media and society. Her research interests include the cultural impact of media, the role of culture in communciation systems, space communication, the cultural ecology of media, and new policymaking paradigms for the 21st century. She has had extensive public and community television experience including research, writing, and production.

Bruce Henderson   (bruce.henderson@colorado.edu) is on the journalism and mass communication faculty at the University of Colorado at Boulder, prior to which he was in professional journalism for 20 years. He has created numerous on-line publications including the *Campus Press*, a student-produced newspaper at the university. He also created the Colorado Press Association Web site, which allowed Colorado papers to create their own Web editions. He has authored several Web computer applications.

Kevin Kawamoto (kawamoto@u.washington.edu) is an assistant professor of communication at the University of Washington. He formerly was the manager of technology studies at the Freedom Forum Media Studies Center in New York. Previously, while a doctoral student in communications, he was a research and teaching assistant at the University of Washington. He has been a degree fellow at the East-West Center in Honolulu for 2 years and was the Crown Prince Akihito Scholar at Nanzan University in Nagoya, Japan. He writes and speaks on technology issues for a number of publications and organizations.

Mark R. Levy (mlevy@jmail.umd.edu) is professor and director of the Center for Research in Public Communication in the College of Journalism at the University of Maryland. He has just completed a 5-year term editing *Journal of Communication*. He has published 10 books and nearly 50 scholarly and professional articles. His most recent co-authored monograph, *Global Newsrooms, Local Audiences: A Study of the Eurovision News Exchange,* examines the globalization of television news and its impact on viewers worldwide.

John E. Newhagen (johnen@umd5.umd.edu) is an associate professor of journalism at the University of Maryland. He worked as a foreign correspondent throughout Central America during the 1980s including a stint as head of UPI's bureau in

El Salvador. He also worked as a regional editor in Mexico City, and foreign editor in Washington, DC. He earned a PhD in communication at Stanford University. His research interests include the psychological effects of emotion-laden television images and the role of new media in journalism.

Susanna Hornig Priest (E341SH@tamvml.tamu.edu) is an associate professor of journalism at Texas A&M University. She holds a PhD in communications from the University of Washington. She has a long-standing interest in the relationship among science, technology, and society from a communications point of view and has taught courses on communication technology and society, media and society, public opinion and the media, and journalistic information-gathering. She is presently on leave from Texas A&M and a staff member of the LEAD Center, College of Engineering, University of Wisconsin-Madison, where she is doing research on the instructional potential of information technology.

Jason Primuth is executive producer of *Channel 4000*, a popular local content site on the Internet. He assists a staff of 15. Earlier he established a consulting group to drag media entities into the interactive world. Yanked out of the ivory tower of George Washington University by CBS, he has been following the trends in information technology for the past few years from Minneapolis.

Steven S. Ross (ssr3@columbia.edu) is associate professor of professional practice at Columbia University's Graduate School of Journalism where he has taught since 1985. He has created a spreadsheet, database, and Internet required course for master's-level students. He also teaches national and environmental reporting and a new media workshop. He also writes a monthly column on computers for *Architectural Record* and continually updates a book on computer technology. He has authored or edited 17 books and 3 commercial software packages. Much of his earlier career was spent as a magazine writer and editor.

Wendy S. Williams (wendyw@american.edu) is an assistant professor and director of the undergraduate writing program in communication at American University from where she earned a master of arts. Prior to 1989, she worked for 7 years as a reporter and assistant editor on the financial desk of *The Washington Post*. She teaches writing, reporting, and media studies courses and has led the School of Communication's efforts to teach computer-assisted reporting as part of the required skills curriculum. She has researched the role of advertisers in influencing editorial decisions on business news. She has written for many national journals including *Washington Journalism Review,* and *Newspaper Research Journal.*

Index